BRAVING
THE
ELEMENTS

Also by David Laskin

A COMMON LIFE: FOUR GENERATIONS
OF AMERICAN LITERARY FRIENDSHIP AND INFLUENCE

EASTERN ISLANDS: ACCESSIBLE ISLANDS
OF THE EAST COAST

THE READING GROUP BOOK

DAVID LASKIN

BRAVING
THE
ELEMENTS

THE STORMY HISTORY
OF
AMERICAN WEATHER

ANCHOR BOOKS
DOUBLEDAY
NEW YORK   LONDON   TORONTO   SYDNEY   AUCKLAND

AN ANCHOR BOOK

PUBLISHED BY DOUBLEDAY

ANCHOR BOOKS, DOUBLEDAY, and the portrayal of an anchor are trademarks of Doubleday, a division of Bantam Doubleday Dell Publishing Group, Inc.

The author gratefully acknowledges permission to quote from the following:

"Fire and Ice" from *The Poetry of Robert Frost* edited by Edward Connery Lathem. Copyright 1951 by Robert Frost. Copyright 1923, © 1969 by Henry Holt and Co., Inc. Reprinted by permission of Henry Holt and Co., Inc.

"The Snow Man" from *Collected Poems* by Wallace Stevens. Copyright 1923 and renewed 1951 by Wallace Stevens. Reprinted by permission of Alfred A. Knopf, Inc.

Book design by Jennifer Ann Daddio

The Library of Congress has cataloged the Doubleday edition as follows:

Laskin, David, 1953–
Braving the elements: the stormy history of American weather / David Laskin. — 1st ed.
p.   cm.
Includes bibliographical references and index.
1. United States—Climate—History.   I. Title.
QC983.L38   1996
551.6973—dc20   95-19851
                                                                                    CIP

ISBN 0-385-46956-X

Printed in the United States of America

First Anchor Books Edition: July 1997

1   3   5   7   9   10   8   6   4   2

TO MY FATHER,

AN EVEN BIGGER WEATHER FANATIC THAN I AM

ACKNOWLEDGMENTS

Considering how vast America's weather is, our community of weather professionals is astonishingly small and wonderfully generous in sharing their time, knowledge, and enthusiasm. I could never have researched or written this book without the help of weather scientists, forecasters, and historians working at universities, in private industry, and especially in the various branches of the National Weather Service. I am especially grateful to Louis Uccellini, director of the Office of Meteorology at the NWS headquarters in Silver Spring, Maryland; James D. Belville, area manager of the Weather Service Forecast Office for Washington, D.C.; Jerry Mahlman, director of the Geophysical Fluid Dynamics Laboratory of NOAA in Princeton, New Jersey; Frank Field at the CBS weather center in New York City; Allan Eustis in the NWS office of industrial meteorology; and Brad Colman, science and operation officer at the Weather Service Forecast Office in Seattle.

The staff of the National Meteorological Center in Camp Springs, Maryland, walked me through step-by-step the process of assembling national forecast guidance from computer-generated forecast models. In particular I would like to thank Dave Olson and Paul Kocin of the NMC's Meteorological Operations Division; James G. Howcroft, the NMC's deputy director; and Kenneth Comba.

Staffers at the National Weather Service headquarters sat cheerfully through lengthy interviews and answered numerous follow-up questions. Thanks go to: Frank Lepore and Bud Littin in the public affairs office; Edward M. Gross in the office of industrial meteorology; Kevin C. McCarthy, operations officer; Ron Gird, satellite program leader; Scott Kroczynski, manager of the hydrometeorological information center; Donald R. Wernly, chief of the warning and forecast branch; Tom Blackburn, manager of the national cooperative program; Eli Jacks; Russell Martin at the Climate Analysis Center; and Linda S. Kremkau for helping me find photos.

Ira Geer and Keith Seitter of the American Meteorological Society fielded many questions and gave me good leads. I'd also like to thank Ronald C. Taylor, program director of physical meteorology at the National Science Foundation, and John Snow, dean of the College of Geosciences at Norman, Oklahoma. Dean Cosgrove shared with me his passion for storm chasing.

David Ludlum, the unofficial dean of American weather history, answered questions, shared research tips, and pointed me in the right direction. I highly recommend his books about national and regional weather history, especially his two volumes on early American winters.

Norbert Untersteiner, Richard Reed, and Robert Fleagle at the University of Washington Department of Atmospheric Sciences helped me get a handle on the weather of the Pacific Northwest. Larry Kierulff, lead forecaster at the Seattle Forecast Office, was also extremely helpful and patient, as were Andy Stern and Jim Weismueller at the Sterling Forecast Office. Jack Weatherford and Ed McGaa talked with me about Native American attitudes toward weather.

I had a far clearer and more sophisticated grasp of the exciting research being done on climate change and global warming after talking to and corresponding with some of the researchers working in this area. I'd like to thank James Hansen and George Tselioudis at NASA/Goddard Institute for Space Studies; David Schimel at the National Center for Atmospheric Research; William Bonner at the University Corporation for Atmospheric Research; Syukuro Manabe at the Geophysical Fluid Dynamics Laboratory of NOAA in Princeton. Michael J. McPhaden at the NOAA/Pacific Marine Environmental Lab in Seattle and Fritz Stahr at the University of Washington's Ocean Physics Department helped me understand ENSO events and research in this area.

I had a great time researching the chapter on television weathercasting. Thanks go to Ray Ban at the Weather Channel; Robert Carnahan, commissioner for professional affairs with the AMS; Ray Boylan at WSOC-TV in Charlotte, North Carolina; Gary Shore at KJRH in Tulsa, Oklahoma; Pam Daale at KMGH in Denver; Tom Skilling at WGN in Chicago; and Bob Ryan at WRC in Washington, D.C. Bob Copeland and Robert Henson helped fill in the historical background. For helping me understand the role of private weather companies I'd like to thank Joseph D'Aleo, chief meteorologist at WSI; Bill Schlueter, vice president for marketing at Kavouras Inc. in Minneapolis; and Mike Smith, president of WeatherData Incorporated in Wichita, Kansas.

I'd like to thank Gary Grice, Diane Boxley, and Mindy Black for helping me with photo research. And I'm grateful to the late poet James Merrill, from whom I borrowed the title of this book. "Everything changes; nothing does," Merrill wrote in his poem "After the Fire," which he included in his *Braving the Elements;* and those four words might stand as a fitting epigraph to *my Braving the Elements.*

I owe a tremendous debt to my agent, the late Diane Cleaver, who passionately believed in this book and who brought her passion to bear in making it happen. Diane instantly "got" the idea, she gently prodded me to sit down and put it on paper, and she worked her usual magic in selling it to Doubleday. I feel blessed that Diane was able to read the complete manuscript before her sudden death in April 1995—and I know she was extremely excited about seeing the book through to press. I, like all of Diane's many friends and clients, deeply mourn her passing. She was unique. I'll think of Diane whenever I hold this volume in my hand. I'm grateful to Heide Lange for picking up where Diane left off and for sharing her own memories of Diane with me.

My editors at Doubleday, Renee Zuckerbrot and Bill Thomas, have been generous with praise and precise with criticism. Thanks also go to Joel Fishman for signing the book up and to Jacqueline LaPierre for help with all the many details of getting a manuscript into print.

Friends and family members have, as always, been a source of encouragement, support, advice, and information. Phil Patton passed along data and contacts whenever he came across them. Special thanks go to my brother Bob, my brother-in-law Lawrence O'Neill, Jr., and my father-in-law, Larry O'Neill, Sr., for answering scores of my dumb questions with patience and clarity. And I want to thank my own father, Meyer Laskin, who sparked my fascination with weather years ago in the Adirondack Mountains when he pointed to a single bright spot in a canopy of sodden gray sky and vowed that it would clear. I've been scanning the skies and religiously tuning in to the weather forecast ever since.

My wife, Kathleen O'Neill, and my daughters, Emily, Sarah, and Alice, have kept my own inner weather bright and sunny during the *Sturm und Drang* of researching and writing this book. I want to thank the girls for accompanying me to the Seattle Forecast Office, and for constantly asking me when my "weather book" would be done—it kept my nose to the grindstone all these years.

CONTENTS

INTRODUCTION

THERE WAS JUST ONE WEEK TO GO UNTIL SPRING AND THE SUN shone brightly over New York City. The air temperature had been rising steadily through the morning, and by midday it hovered in the low 40s. Manhattan office workers on their lunch breaks were walking around with their coats unbuttoned and smiles on their faces, at least some of their faces. Not a lick of wind to cut the sunshine. March 12, 1993. Friday. You could practically taste the weekend.

I heard the blizzard warning early that afternoon while I was browsing in a Fifth Avenue bookstore. A clerk at the cash register had a radio on. The music stopped and those scary, end-of-the-world bleeps came over the air. "This is not a test. This is not a test." *The voice of doom.* "This is a broadcast of the Emergency Broadcast System. A blizzard warning has been issued for Saturday for the New York City metropolitan area, Long Island, New Jersey, and Connecticut. Repeat: A blizzard warning . . ." Gritting my teeth a little, I paid for my books and headed back out into the sunshine. Not a cloud in the sky. Maybe a faint smudge of gray over the Hudson. Blizzard? Nah. No way. Never happen.

Meanwhile, at the National Meteorological Center outside of Washington, D.C.—the forecast high central command for all of North America —the phone was ringing off the hook. At the NMC, the blizzard had already been howling all week—howling through the circuits of cyberspace. The first inkling that some major weather was in the works came on Monday, when one of the NMC's computer-generated numerical models latched on to a suspicious-looking low down in the Gulf of Mexico. By Wednesday, the NMC forecasters were predicting that weekend snowfall would be measured in *feet* inland with eight or more inches accumulating closer to the coast. Now, on Friday afternoon, just ten hours before the storm was supposed to hit, reporters were swarming all over the place. How much was going to fall on Washington? Where exactly was the storm now? Was this really the Big One? At some point amid the buzz and hum of reality metamorphosing into news, the phrase "Storm of the Century" was born. Reporters clamored to get Dr. Louis Uccellini, chief of the NMC's Meteorological Operations Division, to use this phrase, but he wouldn't do it. "It's going to be a hell of a storm," he kept telling them, "but the century's not over yet."

In fact, Uccellini was a nervous wreck. In the latest computer runs, the various models had diverged radically in their predictions of how deep the low would be. Even more alarming was the fact that one dissenting model placed the storm track well inland, which meant rain—not snow— along the coast. Rain in Washington. Rain in Philadelphia. Rain in New York. Rain—even though the blizzard warnings were already bleeping away on millions of radios up and down the eastern corridor. As Friday afternoon deepened into dusk, many cups of coffee congealed next to computer terminals at the NMC, many fingernails got chewed to the nub.

The storm dominated the Friday-evening news even though a single flake had yet to fall on the major East Coast cities. Television reporters clutching microphones stood outside on dry city streets and warned viewers to "brace" for a "monster weather event." Then back to the weather center, where the station's smiling meteorologist was "monitoring" the "situation" moment by moment. In the absence of real-time snow flakes, producers had to resort to stock footage from previous storms: cars stuck in drifts, parka-clad suburbanites shoveling their driveways, rush-hour traffic creeping along slushy highways. "This could be the biggest and meanest storm ever," Jacqueline Adams told the viewers of the CBS evening news. And there was Lou Uccellini, looking nervous, explaining that this storm was going to be a kind of winter hurricane, plunging barometric pressure to record lows along the East Coast. On the Weather Channel you'd think it was the eve of World War III. "You know this one will come through," Washington, D.C., weathercaster John Curley promised his audience.

But would it? Hadn't the National Weather Service been wrong before? Hadn't the NMC, in fact, gotten killed on an East Coast winter storm prediction just two weeks ago? Even as they filled precious airtime hyping the approach of the blizzard, the television reporters were careful to let just the shadow of a doubt fall. If the storm never came, if the snow turned to rain in Virginia, it wouldn't be *their* fault. . . .

When I woke up the next morning, two to three inches of fresh snow covered the earth and more was coming down fast. An occasional gust of wind eddied the flakes. My kids were jumping for joy. So they nailed it after all, I was thinking, strangely surprised and also a little disappointed. Was this really "Armageddon! The mother of all blizzards!" as a CNN reporter yelled at some point during the day? To me it looked like just an old-fashioned snowstorm, the kind that used to make me jump for joy when I was a kid. Frankly, I'd expected something more cataclysmic.

It wasn't until the next day, when I read about the extent of the storm in the newspaper and the damage it inflicted on life and property from Alabama to Maine, that I understood what we had lived through. What had happened "out there" in the atmosphere was, in an odd way, beside the point. What made this "superstorm" *weather* was its impact on us: our lives, our homes and businesses, our technologies, our emotions, our communications systems. It is my conviction that weather doesn't just happen—it happens to us. Around us *and* inside us. Weather is our internal experience of daily atmospheric change. It is a human fabrication—a compound of rainfall and ritual, economics and air pressure, science and superstition, desire and expectation. It begins, naturally, "out there," but the "out there" only becomes weather after it has touched us and we have touched it. "However one experiences it, the weather is personal, as an enemy or friend," writes science journalist Thomas Levenson in his book *Ice Time*. "Its impact on human life is obvious and immediate."[1] We pray for it, curse it, or glory in it; we measure, analyze, and try to predict it; we hide from it, chase it, and die in it. Weather has a history—a strange and complicated history that involves attitudes, beliefs, scientific instruments, government bureaus, poetry, sermons, theories, corporate propaganda, the mass media, and the private thoughts and feelings and actions of millions of ordinary people. Recounting this history as it has blown across the United States from the arrival of the Paleo-Indians in North America to the latest computer forecast model run at the National Meteorological Center outside of Washington is the subject of my book.

IT IS A HISTORY AT ONCE UTTERLY UNREMARKABLE—FOR what could be more commonplace than "the mere weather," as Wallace Stevens called it in a poem?—and stupendously cataclysmic. As Francesca Lyman writes in her book *The Greenhouse Trap,* "Weather is a fact of life. All human beings experience it from their first breath to their last. It has made a difference in battles, love affairs, and sports. But more than with most things, we take its daily variations for granted. It takes weather anomalies—blizzards, tornadoes, hurricanes, or temperature extremes—to remind us that the world is still governed by nature."[2]

A history of weather is a formidable undertaking, but also an intensely engaging one, for this country is blessed and cursed with the most and greatest variety of extreme weather in the world: "A normal year [in the United States] brings hurricanes, blizzards, +100° F heat, −20° F cold, jungle humidity, desert dryness, and the world's most numerous and strongest tornadoes," reports *The USA Today Weather Book.*[3] "In an average year, weather-related events claim the lives of several hundred Americans," a National Oceanic and Atmospheric Administration (NOAA) fact sheet states dryly, "and damage runs in the billions of dollars." All of this severe weather bombards astoundingly diverse groups of people, each with its own peculiar attitudes and beliefs about the weather. "People don't have any memory, meteorologically," remarks weather historian David Ludlum. But they do have highly sensitive weather antennae—a keen inner sense of what the weather "should be" based on years of exposure, folklore, daily small talk, conventional wisdom, science, pseudoscience, and fantasy. Whatever weather you have grown up with becomes "your weather"—the baseline to which you compare all the other weathers you travel through for the rest of your life. Often unconsciously, you look for that old familiar weather and you brand all weather that departs from it strange. In America, with its ceaseless immigration from abroad and internal migration from one region to another, the weather has been strange for centuries. Columbus was amazed by the ferocity of a Caribbean hurricane. The torrid summers and frigid winters of Massachusetts stunned the English colonists who had come from a mild maritime climate. New Englanders in turn were laid flat by the anomalies of weather in the western states—blizzards in the Rockies in October, tornadoes on the Plains in May, years of total drought followed by torrential rains. We still can't get over our weather extremes—think of just the last few years, with record-breaking cold winters in the late 1970s, record-breaking heat and drought during the summer of 1988, Hurricane Andrew, the most expensive weather disaster in history, hitting Florida in August 1992, the Storm of the Century roaring up the East Coast in March 1993, ceaseless rains flooding parched California during the winter of 1994–95, the blizzard of 1996 which kicked off a record-breaking snow season in the Northeast in early January, floods inundating parts of the Pacific Northwest that same winter after record rains soaked western Oregon and Washington and warm temperatures melted heavy mountain snow packs, a drought rivaling the Dust Bowl days of the 1930s settling over Texas, Oklahoma, and southern Kansas in early 1996 and intensifying through the spring and summer, Hurricane Fran blasting through the Carolinas and Virginia in the first week of September 1996, claiming at least thirty-three lives and sending the Potomac River seven feet above flood level in the nation's capital, and even more devastating floods hitting the West Coast states in the final days of 1996 as heavy snow fell over much of Washington state and in the mountains of Oregon and California followed by rapidly rising temperatures and heavy rain from San

Francisco north. And now, on top of all of this, we're concerned that we may be on the verge of unprecedented climate change brought on by a buildup of so-called greenhouse gases. This sense of the strangeness of our weather, the conviction that we're living in an extraordinary time or place climatically, crops up again and again throughout our weather history.

Weather culture, if I can use that term, splinters by region, by religion, by occupation, hobby, and temperament. One person's weather disaster is another's supreme thrill. Rain clouds have a very different meaning to farmers and vacationers—and this meaning has shifted nationally as we have changed from a predominantly rural agricultural people who worked outdoors and rested indoors to a predominantly urban/suburban people who for the most part earn our livings under shelter. Even today, the old attitudes persist wherever weather determines prosperity. Not only does the weather report command more time on television stations in the Midwest and Plains states, it commands more respect. No sensible weathercaster jokes about rain during a drought summer in Kansas or Montana; while in New York, no matter how parched the earth is, they wring their hands over every shower. But there are signs that attitudes are changing even in the cities and suburbs. "Most people didn't give a goddamn about the weather when I started out on television thirty years ago," veteran New York weathercaster Dr. Frank Field told me. "But now with the rise of outdoor sports, summer homes, and weekend recreation, people care a lot more. The weathercaster has become a more visible figure." These days, as the natural world recedes around us, we yearn to be closer to the weather— to be in touch with its daily rhythms and even to witness firsthand its most savage displays. Amateur storm chasers are on the rise, reports a recent piece in *Smithsonian*—people who come to Texas and Oklahoma in spring and early summer to track down tornadoes, "straight line" windstorms (brief, severe storms packing winds of up to 100 miles per hour), and dazzling outbreaks of lightning like big-game hunters on African safaris. Storm chasing became all the rage after the release of the movie *Twister* in the spring of 1996. Suddenly everyone wanted to get in on the act, and reporters had a field day profiling "real" chasers and comparing their adventures in the field with the silver screen heroics of Helen Hunt and Bill Paxton. "It's the fulfillment of a lifelong dream of mine," says Dean Cosgrove of Log Lane Village in northeastern Colorado, who runs a small business chasing severe weather and selling video footage and still photos to television stations. "I love being out in nature, watching the closing of the air masses. I love the freedom of being out on the Pawnee Grassland in the High Plains—you can see forever and there's nothing man-made. The raw energy of weather out there is a kind of spiritual experience." "What you see is tremendous," comments Jeff Rosenfeld, managing editor of *Weatherwise* magazine, of the storm-chasing experience, "it's simply awe-inspiring."

Awe-inspiring if you're getting footage of it to sell to a local television station; awful if it has just destroyed your home, your car, or killed someone you know. Our weather history is rife with disaster stories. Deadly lightning strikes. Starvation winters. The year without a summer. Blizzards and tornadoes in the West. Dust bowls. Hurricanes. Floods. The Storm of the Century. Again and again through the centuries, victims of severe weather compare the experience to combat. They speak of carrying permanent mental scars, horrifying memories they cannot shake off. "We survivors feel small, vulnerable, and hurricane-obsessed," wrote a Miami-based reporter in *The New Yorker* after living through Andrew. "I used to love the blackening rush of the thunderstorms and squalls of high summer. Now I cringe when the winds gust."[4] Survivors of nineteenth-century blizzards on the Plains and pioneers who saw their life's work whirled away in a few seconds of tornado recorded similar feelings. Weather as deity. Weather as demon. These images recur.

Central to our experience of weather is our endeavor to explain and account for it. What makes it rain? How do storms move? Why does deadly weather strike when and where it does? Can we change the weather? People, of course, have been asking these questions from the beginning of time—but the answers we have come up with here in America have taken on a special quality because of the circumstances of this country's founding. Bacon, Galileo, and Newton were setting down the principles of modern science during the period when North America was colonized, and yet religion dominated colonial thought and society—indeed it remains a dominant force in our society today. Religion and science have been the twin poles of American intellectual life right from the start, and their sometimes tense coexistence has always colored our experience of weather. The same bolt of lightning that fell from the sky over eighteenth-century Pennsylvania was divine retribution, a discharge of electricity, or the winking of the thunderbird depending on who witnessed and described it. By the late nineteenth century, government meteorologists were analyzing national data communicated via telegraph to predict the next day's weather with a high degree of accuracy, but that didn't keep people from praying for the kind of weather they wanted—or from interpreting weather as a sign of God's moods or intentions. It's still going on today. "I believe God is fed up with the sins of America," one woman railed to a New York *Times* reporter in the depths of the miserable winter of 1993–94. "I've been in New York for seventeen years. This is the coldest, wettest winter. God is fed up." "I have seen firsthand evidence that people can affect severe storms through prayer," Tulsa weathercaster Gary Shore of station KJRH said to me at the end of our interview. "People here feel they have some power over storms, that God will hear their prayers. Sometimes I look at my Doppler radar and I hope and pray that what I see there doesn't kill anyone." God sent Hurricane Andrew to "bring us together" a woman told *The New Yorker* reporter quoted above. Meanwhile, in Fort Collins, Colorado, Dr. William Gray, a professor of meteorology at Colorado State University, has worked out a formula for predicting how many hurricanes will strike each year and how many will be major. "There's no hocus-pocus here," says Dr. Gray. "It's just straight objectivity."[5]

In the chapters that follow, I trace the evolution and intertwining of these attitudes—God's will, hocus-pocus, and straight objectivity— through the course of our history. Indeed, I attempt to begin even before recorded history by examining some of the weather rituals and myths that Native American people developed before the arrival of the Europeans. Faith has always played a crucial part in our weather—faith that weather would "make sense" if only we could fathom the mind of God, faith that the weather was a token or a sign whose meaning might one day be clear to us, faith that we could eventually predict the weather perfectly, even gain total mastery over it, if only we knew enough about it. "The atmosphere is much too near for dreams," wrote Cleveland Abbe, the first chief forecaster for the fledgling National Weather Service (then known as the Signal Service). "It forces us to action. It is close to us. We are in it and of it. It rouses us both to study and to do. We must know its moods and also its motive forces. We must conquer it in our struggle for existence." The science of meteorology rose on this spirit of conquest—the optimistic conviction that reason and science and action would eventually triumph over nature's mysteries. But a century of science and technology later, the mysteries remain; indeed some scientists have learned to respect and even revere them. David Schimel, a scientist working in the Climate and Global Dynamics Division of the National Center for Atmospheric Research, sees a fundamental dichotomy in contemporary climate science between what he calls the "technocratic element" and a more intuitive approach: "Some say that better answers will come from more data and bigger and faster computers. But another school believes that the best way to understand the element of mystery in nature is through a free exercise of human creativity. I find the latter view more emotionally satisfying."

I believe that it is this human creativity that has led us to embrace, rather than attempt to conquer, the randomness of weather. To allow for the chaos of weather is not to admit defeat but to take a step closer to the truth. James Lovelock in *The Ages of Gaia* describes how delighted he felt when he learned that "real, honest chaos" had been welcomed into the sciences: "Previously it was believed . . . that given enough knowledge (and, in this age, computer power) anything could be predicted. It was a thrill to discover that there was real, honest chaos decently spread around the Universe and to begin to understand why it is impossible in this world ever to predict if it will be raining at some specific place or time. True chaos is there as the counterpart of order."[6] As Edward Lorenz of MIT, one of the key scientists in the field of predictability research, notes, chaos does *not* mean that attempts at prediction are doomed; but it does mean that there is a limit to how well we can predict. Defining that limit is the fundamental question that weather and climate modelers grapple with. To explain how chaos confounds long-range forecasting, modelers always haul out their little joke about the "butterfly effect"—"the notion," as James Gleick describes it in his recent book *Chaos,* "that a butterfly stirring the air today in Peking can transform storm systems next month in New York." The silliness conceals a serious point: In simulating something as complicated as the atmosphere, the smallest unaccounted-for element snowballs until it throws the whole calculation out of whack. It is this infinite, imponderable mystery that makes even our most highly refined numerical forecast models go hopelessly awry when we try to extend them out to ten days or two weeks into the future. The butterfly wing or God's breath. At the very deepest level, science and religion blur.

Weather science also blurs into politics, social policy, economics, and popular "movements" and moods. Americans have been worrying about and theorizing about climate change ever since they cleared off the ancient forests that fringed the East Coast for farms and plantations. At various times in our relatively short history, we have been convinced that our climate was growing warmer, cooler, drier, wetter, more varied, or more violent due to our plowing, lumbering, polluting, sinning, or a combination of them. Whatever weird was happening to the weather, it was our fault. Nowadays, of course, many scientists believe that we're causing catastrophic global warming by dumping vast quantities of greenhouse gases, especially carbon dioxide, into the atmosphere. In the more extreme greenhouse scenarios, earth becomes indistinguishable from hell within a century. True, the greenhouse theory of global warming has far more hard science behind it than any of our previous notions about climate change. This is undeniable and important. But one can't help noticing as one reels back through history that we have voiced these same climatic anxieties before, often in remarkably similar language. Environmental guilt and fear of retribution haunt our public discussion of greenhouse warming. Even the parameters of analysis have a kind of end-of-the-world, end-of-the-millennium aura: modelers tick off the dwindling years that remain until we reach the threshold of catastrophe—the doubling of atmospheric carbon dioxide from preindustrial levels—just as religious cultists count down to the year 2000 as the onset of the apocalypse. Millenarians and

climatologists make strange bedfellows indeed—but both parties, at least
at their outer fringes, are heavily invested in the doomsday business.

WEATHER HAPPENS TO EVERYONE—BUT FOR SOME PEOPLE IT
becomes a passion. "I just love weather." I heard this same confession from
nearly every meteorologist or weathercaster I interviewed. Each one remembered the exact moment, usually early in childhood, when weather
began to dominate his or her life: a tornado loomed on the horizon; a
blizzard buried their town; a hurricane snapped the tree outside their
bedroom window. From then on, they felt compelled to keep track of the
weather, figure out how it worked, learn how to measure and predict it.
The self-described "weather nuts" who staff the National Weather Service
and the weather centers of television stations and private forecasting outfits
are in very good company, for some of our most distinguished political and
intellectual leaders have been weather nuts too. George Washington jotted
down daily observations of the weather. Thomas Jefferson not only kept a
meticulous weather diary, but urged others around the country to do the
same in the hopes of forming a nationwide network of weather observers.
Jefferson's rebuttal of the so-called degeneracy theory (a French-led attack
on the wholesomeness of America's climate) is one of the great documents
of colonial American science. Ben Franklin, of course, risked his life in his
famous experiment to determine the nature of lightning; equally ingenious
was his work tracking the movement of major storms up the East Coast.
Joseph Henry, the first director of the Smithsonian Institution, made
weather one of the priorities of the new organization and finally realized
Jefferson's dream of a national network of weather observers. John Wesley
Powell carried out the first scientific analysis of the climate of the arid
regions of the West.

   In describing the weather work of these and other illustrious pioneers, I found that theory was impossible to disentangle from technology.
The invention of the telegraph coincided with the emergence of meteorology as a serious science. The airplane added a crucial new dimension to

weather science, giving meteorologists access to the upper atmosphere and leading to the discovery of the jet stream, a key element in medium- and long-range forecasting. The almost simultaneous development of television and the electronic computer has revolutionized weather in the past fifty years. Computers execute the billions of calculations required to generate the numerical models that form the backbone of contemporary forecasting; they create the graphics that appear on television weather reports; they store and sort and transmit weather data. Television has had an incalculable impact on our national experience of weather. For millions of people, television weather *is* weather: surveys have shown that what viewers *most* want to see on television weathercasts is not, as one might expect, the forecast, but rather the summary of current conditions. They want to know what's happening outside their houses *right now* as they watch, one weathercaster told me, and they'd rather get it from the screen than look out the window or go out and put their faces in it. Somehow the weather is more real at one video remove.

Or maybe it's just more amusing. Over the years, television has gone all out to make weather fun or funny or bizarre or outrageous, with "weather girls" strumming ukuleles and drawing cartoons in the 1950s, "happy talk" weather anchors whipping up yet another go-go-go-gorgeous day in the 1970s, and Willard Scott still riding high on the *Today* show as he keeps us all guessing about how much hair he'll be wearing on any given morning. These characters are weather nuts of a different variety— but they're as much a part of the history as Ben Franklin or Joseph Henry or Cleveland Abbe. Less distinguished, maybe, but more intimately a part of our lives. "I feel naked if I go out the door in the morning without watching the television weather report," one meteorological enthusiast I know told me. There are lots and lots of people out there (present company included) for whom the weather is the single most important segment of the news.

Thomas Jefferson rose at dawn each day to check the temperature, sky conditions, atmospheric pressure, and wind direction. Nowadays we get the same information round-the-clock from the Weather Channel, or on-line through our home or office computers, or through the Internet, or from NOAA weather radio, or, for the true devotees, from a home weather station that gets wired directly into a personal computer: flick the switch, hit the button, and watch your very own video display terminal bloom with a color chart of temperature, wind speed and direction, windchill, barometric pressure, and humidity readings; boot up your modem and you too can scroll through the latest National Weather Service maps, satellite imagery, radar, and zone forecasts for the entire country. Being a weather fanatic has never been easier or more rewarding. And it just keeps getting better. As sources of weather information proliferate, weather reports will become increasingly zone- and time-specific. The day is coming when we'll be able to flip open our cellular phones or computer notebooks and find out what's going to happen weatherwise within our zip code during the next three hours. Our VDTs will be vibrating with interactive, fly-through simulations of weather systems—click the mouse if you want to zoom up to the 500-millibar height patterns; click again for all precip within a 125-mile radius; click again to hitch a ride on that tornado. . . .

Meanwhile, we'll still be praying or dancing for rain, theorizing about climate change, and cursing "the weatherman" when those next three hours turn out to be less spectacular than "he" promised us.

It's fitting that weather, the classic example of chaos at work, should have spawned a fitful, nonlinear, multilayered history, a history of jarring juxtapositions and unexpected veerings, of diversity and simultaneity, of great overarching patterns and senseless scramblings. In the history of weather, superstition flies in the face of science and science keeps swinging beyond the range of the imagination's radar. I was amused to come across a recent piece in the New York *Times* Travel section called "Confessions of a Rain Goddess," in which novelist Rachel Billington insists that her mere presence in any spot is enough to bring on torrential rains: "I once went to the Sahara Desert in search of sun. . . . [I]t rained so hard that flowers popped out of the sand and the Berbers suggested keeping me there as their personal rain goddess . . . I had joined the least desirable club in the world: those who bring bad weather to vacations." I suspect many people, including those with deep faith in meteorology, carry around their own personal little weather mythologies, biases, preferences, eccentric quirks, and odd theories. My own idiosyncrasy is a suspicion, no, an anxiety that whatever weather pattern we're in the midst of will never shift: rain becomes unimaginable to me in a dry July; a sodden March will never warm to spring. And I grow irascible when I don't get the kind of weather I want: to me, a warm spell in winter is akin to a stock market collapse or the defeat of my preferred candidate. "My bias is I like cold weather," Brad Colman, the science and operations officer at the National Weather Service forecast office in Seattle, admitted when I asked his personal opinion of global warming. "It doesn't snow enough in Seattle as it is. So I don't want warming." Not quite an answer to my question, but illuminating nonetheless. "I'm a nasty-weather freak," begins another article in my clipping file. "Hurricanes . . . downright fascinate me. Drenching rains are my nirvana; heaping snowstorms a feast for my eyes."[7]

Our weather history is studded with these kinds of confessions, crotchets, laments, pet peeves. The effect of weather on mind and body has engendered fascinating subcategories of literature, science, and rumination. A century and a half ago, Herman Melville wrote in the novel *White-Jacket,* "[T]here are many people, naturally gay and light-hearted, who, whenever the autumnal wind begins to bluster round the corners, and roar along the chimney-stacks, straight become cross, petulant, and irritable." Now we call this weather-related depression "seasonal affective disorder"—SAD—and psychologists have found that it affects millions of people every winter, especially in the northern latitudes. Our poets write of inner weather ("One must have a mind of winter / To regard the frost and the boughs / Of the pine-trees crusted with snow . . ." Wallace Stevens begins his poem "The Snow Man"). And weather crops up again and again in our national literature as metaphor, symbol, spiritual force, expression of the eternal mysteries of nature. James Galvin describes a kind of mystical yet practical merging of weather and character in his Rocky Mountain novel *The Meadow:* "Lyle . . . didn't have moods, he had weather. Not some inner weather that could have been a mood—Lyle had *the* weather. Inside him he had going on exactly what was going on in the sky, or some combination of recent weather and what was likely to develop . . . That high in the mountains a man lives less on the land than in the sky. After forty years the weather had all the bearing." Such notions swirl through our diaries, fictions, newspaper articles, daily conversations. The trick is always to tease them out, assemble them, make some sense of them down through the centuries and out across the continent's broad back.

WEATHER IS THE GREATEST EVENT ON EARTH AND THE MOST reliable: it will be there tomorrow, no matter what. It is something that we all have in common, even if each of us experiences it very differently. It is our essential link to the natural world and, along with landscape, the prime quality that defines the sense of place. It is arguable that weather, more than geography, determines the character of America's regions: Northeast, Southeast, Midwest, Great Plains, Southwest, Great Basin, Northwest, Pacific Coast—each with its own set of weather variations and a population attuned to these variations and shaped by them. When we move (as I did, in the course of writing this book, from New York to Seattle) weather is the last thing we leave behind and the first thing we find when we arrive. Weather, in a sense, *is* home. And yet, for many of us, for long stretches of time, it scarcely exists. Quick: Without consulting a newspaper, can you recall what the weather was like yesterday? The day before? The day before that? And what, really, does it matter? The beautiful days come and go as regularly, as inevitably as the stormy ones. We live most of our "real" lives indoors anyway. Weather has become decoration, a movie that plays against the picture windows of our homes. And then, without notice, the weather rears up and slams through our lives and communities like a bomb or an epidemic. More than 2,000 people dead in the Johnstown, Pennsylvania, flash floods of 1889. Hurricanes Camille, Eloise, Gilbert, Hugo, and Andrew killing hundreds of people, and destroying billions of dollars' worth of property just in the past three decades. The dust bowl of the 1930s. The long winter on the Great Plains in 1880–81. The blizzard of 1888 and the Storm of the Century in 1993.

Events like these become public and private benchmarks dividing time into before and after. As Levenson writes in *Ice Time,* they "form the landmark around which we hang the memories of a particular time and place."[8]

We are constantly making and revising the history of weather, but weather itself is ahistorical. Infinite, fathomless, incalculable, it just keeps happening, regardless, every day. It lives in the eternal present tense, forever new and of the moment, always wiping the slate clean and starting all over again. It is immune to our human agency, desire, moods, memories, theories, and it has proven itself to be fairly impregnable to our attempts to change or control it. A history of this stupendous force is thus doomed from the start to incompleteness, for weather never stays still long enough to have its portrait painted: the best we can hope for is to freeze some of its most dramatic gestures in snapshots.

This book is essentially a collection of the best snapshots I could get, beginning with the role of weather in luring Asian nomads across the Bering Strait land bridge during the last great Ice Age and extending through every major wave of exploration, settlement, discovery and rediscovery, social and cultural and technological shift. Through the years, treatises, proverbs, folk sayings, sermons, editorials, scientific theories, and personal narratives furnish many telling juxtapositions. The history of forecasting in America, the evolution of the National Weather Service out of a network of weather observers set up through the Smithsonian Institution, the twists and turns of television weathercasting, the fierce current debate over global warming—all are part of the story. Ever since humans arrived here, America's weather has been a source of bafflement, awe, illusion, dashed hopes, and often intense suffering. In telling this story, I have tried to balance theory and emotion, examining the attitudes and assumptions behind weather science and exploring in turn how the increasing sophistication of that science alters our daily private experience of weather.

AS I WRITE THESE WORDS, THE WIND IS BLOWING HARD AND from the south, and thick gray clouds are massing over Puget Sound, raising the hope for some much-needed rain (yes, the notoriously wet Pacific Northwest has had below average rainfall for the past few years, possibly the result of a persistent ENSO event—see page 218—that has shifted global weather patterns since 1991). Meanwhile, at least fifteen people have died in floods in East Texas after more than thirty inches of rain fell this week—another possible consequence of the unusually persistent ENSO. New York, I note, is rather balmy for this time of year, with highs in the upper 60s, though showers are expected. Los Angeles, as usual, is sunny and warm. The early snow that piled up in the Rockies last week is melting. In short, the weather goes on. To me, this daily drama is endlessly absorbing. "There's something different every day," one of the meteorologists who works at the Seattle National Weather Service forecast office told me. "Even when there's nothing much going on, there's still the challenge of trying to get the day's high temperature right." I know just what he means. Like the professional forecasters with their numerical models and the storm chasers with their battered video cameras, I am awed by weather. Its beauty. Its power. Its infinite variability. I love to contemplate the great air masses spiraling off the waves in the jet stream in clockwise or counterclockwise whirlpools. I admire the elegance and ingenuity of the daily weather map with its isobaric patterns, like atmospheric fingerprints, defining high- and low-pressure systems, its saw-teeth cold fronts and round-teeth warm fronts, and its shaded bands of precipitation. I am open-mouthed with wonder at the fact that the supercell of a single intense thunderstorm can pack more energy than a hydrogen bomb.

For me, the best part of writing this book has been my total immersion in weather, its daily phenomena, its strange beauty, the art and science it has engendered, the changes—sometimes trivial, sometimes earth-shattering—it has wrought in people's lives. Native Americans scoff at the notion that America was "discovered," but in a sense we have never ceased to discover it: the land and its weather perpetually take us by surprise, opening our eyes in wonder, forcing us to reexamine not only what we see but how we see. This book is really an endeavor in revision in the literal sense of seeing again: I have tried to look long and hard at weather, to see it through the eyes of others down through the generations of our history.

In searching for patterns in this history, in recognizing chaos where patterns fail to form, I have learned to look at weather in a new light, or rather many different lights at once. To see what's going on "out there" as if for the first time: this has been my inspiration in writing the book. I only hope that I have succeeded in sharing this inspiration in the pages that follow.

1.

NATIVE AMERICAN WEATHER

THE HUMAN SPECIES WAS BORNE INTO AMERICA ON A WAVE OF cataclysmic climatic change. For millions of years, our human ancestors evolved and spread through the Old World—emerging in Africa some two to three million years ago, then migrating into the Middle East, north into Europe, and east to Asia—while the two continents of the New World remained inviolable, sealed off from the transforming touch of man. Then the earth, of its own accord, breached the seal of the Americas. About 100,000 years ago, global temperatures once again started to fall and the ice cap that crusts the North Pole began its fourth radial advance southward, gradually but relentlessly swallowing the tundra of northern Canada, crushing the northern landscapes beneath an enormous wall of ice, chilling and paralyzing the oceans. As more and more of the world's water turned to ice, ocean levels dropped—sinking more than 280 feet beneath their present maximum at the height of the glaciation—and huge shelves of dry, barren ground emerged at the margins of the North American continent. At the extreme northwest of the landmass, where the fifty-odd-mile Bering Strait now separates Alaska's Seward Peninsula from Siberia's Chukchi

Peninsula, a broad, flat land bridge became exposed to the frigid glacial winds. For 41,000 years during the last glacial period—from about 75,000 to 45,000 years ago and again from about 25,000 to 14,000 years ago, when melting glacial ice began to resubmerge the bridge for all time—this land bridge joined North America and Asia as continental Siamese twins.

Beringia, as scientists call the bridge region, was not a hospitable environment during the last Ice Age. Frigid during its nine-month winter, fogbound during its brief summer, Beringia was starved for precipitation and swept by relentless winds generated by the glaciers. Polar desert conditions probably kept Beringia and northern Alaska unglaciated but barren. The low, smooth bridge leading to the unbroken mountain range of North America's western coast was probably a grass steppe—"a featureless but slightly undulating plain," as one writer describes it, "studded with bogs and swamps, and covered by tundra vegetation of sedges, grass and low shrubs."[1] And yet, despite its climatic and geographic rigors, Beringia attracted a wide variety and rich abundance of advanced life forms. Mastodons and woolly mammoths trekked east from Asia across the bridge into North America; horses and camels crossed the other way, from New World to Old. And at some point, perhaps 14,000 years ago, perhaps 18,000, perhaps 40,000, perhaps far, far earlier, *Homo sapiens sapiens*—fully evolved modern humans—joined the traffic on the Beringia bridge.[2] The first pioneers were undoubtedly hunters in pursuit of big game. They came with no plan, no mission, no knowledge of the vast expanses of virgin terrain stretching away to the south and east. As Philip Kopper writes in *The Smithsonian Book of North American Indians,* "There is no reason to believe that these people knew where they were going—or even that they knew they were going anywhere."[3] The migration eastward was not an isolated phenomenon but rather a part of the global explosion in human settlements that began some 35,000 years ago. Seemingly overnight in geologic time, the human race took a giant step forward in technology, sophistication, communication, and the ability to adapt to and dominate harsh environments. The Ice Age, far from stunting human development, seems to have stimulated and challenged our species to advance. We began to wear clothes, to hunt with more precision, to make more efficient use of the game we killed. For the first time, people were *able* to survive the rigors of the Arctic, and where we could survive we seemed determined to explore. The spirit of adventure may well have pushed people beyond the known horizon back then, just as it has done in the succeeding millennia.

In any case, even discounting anything so intangible as the spirit of adventure, there was plenty of game to lure people into and across Beringia. Scientists speculate that the human crossing into America was molasseslike: an imperceptible flow eastward as sons hunted just beyond the range of their fathers, with each generation pushed a few steps closer to the New World by the animals they pursued and by the pressure of the growing settlements behind them. As Nigel Davies writes, "It seems that people just seeped into America across Beringia if the passage was open."

Unlike the European discovery and colonization of the Americas, which was accomplished in a few swift centuries, the original penetration probably took thousands of years. Scientists are still far from resolving the fundamental questions of when these first immigrants set foot on American soil, what part of Asia they originated from, and whether they came in a single slow-moving current or in a series of separate waves. We do know for a certainty that our species was established here by 12,000 years ago, for this is when the rising ocean once more drowned Beringia and severed the link between Asia and America. The Asian nomads who had unwittingly crossed over into the New World were now trapped here. Ten millennia before the greenhouse effect, global warming created America as we know it today and isolated the first Americans here.

It is somehow fitting that the history of American weather—the history of man's experience of the climate of the New World—should have begun with climatic upheaval.

ONCE PEOPLE WERE ESTABLISHED IN THE NEW WORLD, THEY spread fairly quickly throughout the two continents, adapting with astonishing ease to climates ranging from polar to tropical. Within 2,500 years of their arrival—and possibly far less—the Paleo-Indians had reached Tierra del Fuego, planting settlements along the way in mountains, deserts, arid plains, and tropical rain forests. In the waning centuries of the Ice Age, North American climates and topography were considerably different from what they are today. The Midwest had a more uniform climate than at present: summers were cool and windy but winters were probably milder than they are now since the glaciers kept the coldest air trapped up in the Arctic basin during the winter months. The Pacific Northwest and Florida were quite dry at this time, but the Great Basin and the Southwest were considerably wetter and cooler than today due to a southward displacement of the westerly jet stream.[4] In the Great Basin, vast inland seas —including Lake Bonneville, the precursor to Utah's Great Salt Lake, and Lake Lahontan, the precursor to Nevada's Pyramid Lake—filled low-lying areas and turned the mountaintops into islands: at its maximum, Lake Bonneville was as large as Lake Michigan and hundreds of feet deep. Immense marshes at the fringes of these freshwater lakes provided habitat for ducks, geese, and other game birds. The Appalachian Mountains of the middle Atlantic region supported tundra-type vegetation at this time, while spruce-pine forests covered the lower foothills. Tundra was also widespread in northern New England, and the spruce forest that is presently characteristic of northern Maine and the mountains of New Hampshire and upper New York State stretched all the way west across the prairies and plains to the conifer forests of the Rockies, covering much of Iowa and northeastern Kansas, as well as Nebraska and the Dakotas.[5] The prevailing late-glacial climatic conditions of North America would "favor or even enforce a hunting economy," as one anthropologist notes[6]—an assumption that is richly supported by the thousands of arrowheads that archaeologists have unearthed all over the continent near remains of such Ice Age megafauna as caribou, mammoths, musk oxen, long-horned bison, camelids, horses, tapirs, dire wolves, and giant armadillos.

The Ice Age ended even more abruptly than it had begun. Around 12,000 to 10,000 years ago, temperatures started to rise sharply and precipitation decreased over most of the country. Ice retreated steadily northward, releasing at its southern boundaries huge moraines of gravel and rock that it had scooped up during the millennia of its advances. New York State's Long Island and the large Atlantic Ocean islands of Martha's

Vineyard, Nantucket, and Block Island are all composed of such glacial deposits. The marked postglacial warming trend transformed the forests of Wisconsin and Minnesota from spruce to pine in just a few decades, and the sharp curtailment of rainfall allowed the grasslands of the Plains to expand eastward and northward until they eventually covered 750,000 square miles of America's heartland. With the return of the jet stream northward, the Southwest became arid once more and the Pacific Northwest began to receive enough rain for towering forests to replace the tundra and subalpine parkland of the previous millennia.

The Paleo-Indian settlers adapted to these swift climatic changes, and in many regions even benefited from them. The expansion of the prairies opened new grazing lands for the large herbivores—mammoths, bison, horses, and antelopes—and Paleo-Indian hunters followed where the animals led. The boon to the animals of increased grazing lands proved fleeting, however, for around 11,000 years ago most of the Ice Age megafauna began to vanish into extinction. Scientists still do not know whether the mammoths, mastodons, and other North American big-game species were hunted to extinction by increasingly sophisticated and populous Indian groups or whether they perished from an inability to adapt to the postglacial climatic swings. Of the Ice Age megafauna that had once roamed the continent, only the bison herds remained, forming the basis for a human culture that persisted from central Canada to the Gulf of Mexico for over ten thousand years.

Some anthropologists believe that the explosive diversity of Native American cultures can be directly traced to the radically new climatic conditions that followed the glacial retreat.[7] In a comparatively short time span, Indians witnessed the passing of the Ice Age megafauna, the drying up of the lakes and rivers of the Southwest and the retreat of forests up the mountain slopes, the steady rise of the ocean waters over the continental margins, the growth of deciduous forests in the eastern sections of the country and pine-oak forests in the Southeast, the dominance of huge conifers in the forests of the Pacific Northwest and the spread of chaparral over the warm, dry hills of California. The new environments that resulted from climate change forced humans to diversify their means of subsistence

—to supplement hunting with fishing, gathering grains, nuts, fruits, and vegetables, and eventually, in some areas, with farming. Ultimately, the drier, milder weather that settled over North America after the glacial retreat proved to be a blessing to the first Americans. The warm millennium that followed the Ice Age "appears to have been the critical moment when human populations started to grow more rapidly," notes Brian M. Fagan in *The Great Journey.* "The stage was set for rapid cultural, economic, and social change in many areas, change that was still climaxing as European explorers and colonists arrived in the New World."[8]

AS SOON AS THEY ARRIVED IN THE NEW WORLD, THESE EUROpean explorers and colonists commenced to obliterate Native Americans and their cultures—an endeavor that proceeded without interruption for the better part of three centuries. Only in the past century or so have whites attempted seriously to reconstruct what they had so energetically destroyed. Research into the history and culture of pre-Columbian Native Americans is now a major academic industry in the New World—or rather several related industries. The fact that the Indians north of Mexico left no written records has only excited the ingenuity of the academic armies. Archaeologists and anthropologists have worked wonders teasing Indian prehistory out of heaps of earth, stone, bone, wood, and pottery. Their efforts have yielded an immense panorama of thousands of years of cultural succession and paths of influence radiating and intersecting across the continent: the so-called Clovis hunters with their distinctive fluted arrowheads giving way to the Folsom bison hunters around 8000 B.C.; agriculture spreading north from Mexico into Arizona and New Mexico in the millennium before Christ and then into the Southeast and Middle Atlantic states; a materialistic and hierarchical society of fishermen and wood carvers living in comparative luxury in the temperate, salmon-rich Pacific Northwest from around 400 B.C. on; a series of sophisticated pueblo-building cultures flourishing in the dry Southwest while in the south-central states there emerged the powerful Mississippian culture, characterized by large towns, successful farms, and elaborate temple mounds—a cultural pattern that spread from about 1000 A.D. to 1400 over a huge geographical area encompassing the entire Southeast from the Atlantic to the eastern Plains and much of the Midwest. The astonishing diversity of prehistoric Indian cultures is reflected in their languages—some 200 mutually unintelligible languages developing in North America alone.[9]

One fascinating by-product of this recent explosion of information on America's prehistoric Indians has been an increasingly detailed picture of the climates they lived in. Techniques devised to analyze and date Indian sites have also, sometimes unexpectedly, yielded extremely detailed data about rainfall, temperature, and regional climatic swings. Perhaps the best example of this is the development of dendrochronology—a method of determining dates through the study of annual tree rings. A. E. Douglass, an early-twentieth-century Harvard astronomer, perfected this technique in the course of his quest to correlate sunspot activity with weather phenomena. Douglass fixed his sights on the conifers of the Southwest because of their extreme antiquity—stunted ponderosa and bristlecone pines and Douglas firs survive on the region's dry mountainsides for hundreds and even thousands of years. Though he never conclusively proved that sunspots cause climatic change, Douglass did discover in 1929 that he could assign precise dates to the timbers of prehistoric New Mexican pueblos by matching up their tree rings with cores taken from thousand-year-old living conifers. Since tree rings vary in width depending on climate conditions, Douglass could determine the exact year in which a Pueblo beam was cut by comparing its outermost ring with a sample taken from a living tree specimen in the same area. The science of dendrochronology subsequently sprouted the offshoot science of dendroclimatology—a method of reconstructing yearly temperature and rainfall by tree rings. Douglass's successor Harold C. Fritts, working out of the University of Arizona's dendrochronology lab, has married tree-ring analysis to computer technology to provide a detailed record of temperatures, rainfall, and even patterns of atmospheric pressure systems going back hundreds of years. Fritts and his colleagues can't yet tell us whether or not it was raining in Flagstaff, Arizona, on June 1, 1102—but they do know that Flagstaff enjoyed a relatively damp and cool period in the early twelfth century followed by a plunge into drought. Other techniques that climatologists have borrowed from (or shared with) archaeologists and geologists in reconstructing prehistoric climates include palynology (the study of fossilized pollen), analysis of plant fragments and animal bones recovered at prehistoric sites, examination of the contents of pack-rat dens over time, the charting of strand lines and sediments on fossil lakes, and the study of the movements and shapes of sand dunes.

Drawing on data generated by these and a variety of other methods, climatologists have painted in broad-brush the major climatic shifts in the millennia between the retreat of the glaciers and the coming of the Europeans. Although the climates of the United States stabilized at approximately their present levels some 10,000 years ago, there have been significant regional fluctuations over the centuries—significant enough to have an impact on human survival. Most researchers agree that the West, and particularly the Great Basin region bracketed by the Rockies and the Sierra Nevada, went through a cooler and moister episode from about 11,000 years ago to about 7,000 years ago, after which the climate turned a great deal hotter and drier. This long hot spell—known as the Altithermal— endured for some 2,500 years, from about 7,000 years ago to about 4,500 years ago, with temperatures and drought peaking around 6,300 years before the present. After 4,500 years ago, temperature and rainfall levels settled into patterns similar to what they are today, with a gradual trend toward increasing desiccation. The Altithermal probably affected the Great Plains as well, for there is evidence that bison herds were severely reduced or forced out of the region during this period and that lake levels dropped by as much as sixty feet. The Northeast enjoyed a 600-year mild period, from about 300 B.C. to 300 A.D., as indicated by a dramatic rise in sea levels during this period. Then from 300 or 400 A.D. to 800 or 900 A.D., climate conditions changed, with cooler temperatures and not only less annual rain but a shift in the rainy seasons from spring to late summer.[10]

How (and whether) these climatic swings affected the prehistoric Indian peoples is still far from clear. A generation ago, anthropologists and archaeologists believed that the Altithermal rendered the Great Basin and the Plains states too hot and too dry for human habitation, but current researchers are not so sure. The number of prehistoric sites does indeed taper off during the hot millennia (or archaeologists have failed to locate them) and increases markedly after 2500 B.C.; but many now believe that it's possible that Indians did not abandon the region altogether but rather migrated to more favored locations—higher elevations that received slightly more rainfall, for example, or river valleys. In the past, as today, the geographically diverse West abounded in microclimates, isolated weather phenomena, extremely localized vagaries. Bands of showers circulating north into the Plains from the Gulf of Mexico can drench one county and leave the rest of the state dry. The mountains of Utah and Nevada redirect frontal systems in ways that defy prediction. Two and a half millennia of heat and drought sounds catastrophic—and it probably was catastrophic in some areas; but there were undoubtedly tremendous regional climatic variations during the Altithermal, and the Indians undoubtedly exploited these variations to the fullest extent possible.

Our knowledge of the impact of climate on Indian cultures grows sketchier outside of the desert West. Archaic Indian groups in the Northeast flourished during a long warm spell that set in some five thousand years ago, during which nut-bearing hickory trees moved into southern New England and salmon and swordfish became plentiful in the oceans; cooler conditions during the first millennium B.C. probably reduced native populations as pine and spruce took over the hickory forests and clams replaced the swordfish in the Gulf of Maine. The so-called Hopewell Interaction Sphere that emerged in the Midwest around 2,400 years ago—a widespread culture characterized by elaborate burial mounds and exquisite artistry in copper and mica—may have benefited from the mild conditions that prevailed from 300 B.C. to 300 A.D., and the Hopewell decline may have been hastened by climatic deterioration that set in after 300 A.D. Similarly, the rise of the so-called Mississippian culture in the Southeast may have been spurred by the warmer, drier conditions that settled over the region after about 750 A.D. The first really efficient farming took hold in the Northeast between 800 and 1300 A.D., a period in which the climate smiled on agriculture. Agriculture also flourished in the middle Missouri Valley during this balmy period, with fields of maize supporting large villages in Missouri and the central Plains states until around 1100. Then a cool, dry interval after 1250 reduced crop yield, and the larger villages broke up as people migrated eastward.[11] The drought was probably even more severe on the western Plains, where village-dwelling Indians hunted bison and deer during the winter and grew maize in the summer. Archaeologists speculate that mass migrations after 1200 may have emptied Indian villages in Iowa, Nebraska, and South Dakota in a single generation. Farming villages had pretty much vanished from the Plains by the time the first Europeans penetrated the region.

   If one digs deeply enough through the vast literature on prehistoric Native Americans while keeping one eye fixed on the graphs and weather maps assembled by dendroclimatologists, one can piece together at least sections of a puzzle depicting climate change and its impact on population growth, agriculture, migration, and crafts. A warming trend in the Pacific Northwest may correspond with a burst of wood carving and elaborate ceremonialism. A long dry spell in California seems to have put pressure on the wealthy coastal tribes. A series of severely cold New England winters weakened certain groups of Penobscot Indians, possibly opening them to attack from more powerful neighbors. Such academic exercises, however, tell us little more than we can surmise from common sense, and afford little scope for the imagination. The graphs of paleoclimatologists and the shards of archaeologists tell us little about *weather*—man's moment-to-moment, season-to-season experience of the climate in which he is submerged. Even when one is standing beside a prehistoric Indian site in western Pennsylvania or central Missouri, one has a hard time picturing these early farmers harvesting corn in the golden haze of late September or trekking in the direction of the sunrise after a succession of chilly, rainless summers.

IT IS DIFFERENT IN THE DESERT OF THE SOUTHWEST. IN THE Southwest, the force of prehistoric climates seems to linger in the air, preserved like bone and wood and dust by the absence of moisture. Perhaps it is the nature of the prehistoric ruins there—the walls of sandstone masonry mosaic looming out of the canyon floor at Chaco or the toylike blocks of the cliff dwellings hidden for centuries at Mesa Verde. Or perhaps it is the supernatural, inhuman, almost antihuman beauty of the landscapes. In the Southwest, the imagination locks on the confluence of climate and culture. One gazes, withering, at a so-called riverbed devoid of even a whisper of water, and one wonders how and why an advanced agricultural society flourished here for hundreds of years. One contemplates the daily acts of survival in a landscape that suffers the violent wrench of seasonal change in a single day—from icy winter midnight to searing summer noon in twelve hours. The questions of climate—and thus of weather—take on a special urgency in the Southwest. And because of this urgency—and because the urgency echoes so loudly in the abandoned ruins and still resounds in the living cultures of the Pueblo Indians who have inherited the land—scientists have probed the questions of climate more deeply in the Southwest than in any other American region. The questions proliferate, but clear answers remain elusive.

Chaco Canyon today is a stark, forbidding landscape—a sagebrush and juniper scrub desert without a shade tree or year-round water source as far as the eye can see. The opposing sandstone cliffs that form the north and south walls of the canyon are so far apart that one has no feeling of being down in a sheltered, wind- and water-carved declivity. The bottom of Chaco Canyon feels like a wasteland—an oven wasteland in summer and a frigid desolation in winter. An environment best suited to scorpions, lizards, coyotes—not man. It was no different 1,000 years ago, from 900 A.D. to around 1200 A.D., when Anasazi Indians made this the economic, spiritual, and ceremonial center of a unified culture area the size of Scotland. On the floor and the rim of the canyon, the Anasazi erected thirteen elaborate pueblo villages; the largest—called Pueblo Bonito—a vast, D-shaped honeycomb containing 800 rooms, was the most extensive structure north of Mexico. Fifty million pieces of sandstone and beams from an estimated 215,000 trees went into the larger pueblo villages—the stones quarried by hand from the canyon walls and the wooden beams cut from forests as far as fifty miles away and transported to the canyon without pack animals or the wheel. Why did the Anasazi build their capital here? And why, after devoting generations to the construction of Pueblo Bonito, did they abandon it and the other "great houses" of the canyon after scarcely more than a century of habitation? Climate change may hold part of the answer.

Dendrochronology has revealed beyond dispute that the great construction projects in Chaco Canyon coincided with more favorable—that is, relatively wetter—climatic periods. "From A.D. 1050–1130, the heyday of the Chacoan system, precipitation was generally above normal the whole time, except for a sharp but temporary dip in the early 1090s," notes one authority.[12] The Chacoans, like their Hohokam contemporaries in the Southwest, depended on agriculture for survival, making the deserts bear corn, beans, and squash. But where the Hohokams constructed intricate irrigation canals to funnel off water from the Gila and Salt rivers south of Phoenix, the Chacoans were dry farmers who employed an ingenious system of basins, check dams, and chutes to trap, store, and channel the water that fell from the sky. When no rain came or came at the wrong time of year, the Chacoans suffered. Dry farming is a perilous business in the Southwest today, and it was perilous 1,000 years ago. Summer rains tend to arrive in pounding thunderstorms, with torrents that wash away topsoil instead of sinking into it. Growing seasons are short in the upland desert, with late or early frosts frequently destroying the crops. Dry spells were common even during the relatively favorable period from 650 to 1130, though seldom did a drought continue for more than two years. Then, between 1130 and 1349, dry spells began to lengthen, enduring for two and a half to three years. Rainfall patterns began to shift, with winters lengthening and more rain coming in the winter months, which caused flooding in the spring. From 1275 to 1300 rainfall dwindled to the vanishing point throughout the Southwest, bringing on one of the worst droughts of the past five thousand years.

Drought alone, however, cannot explain the decline of the Chacoan system and the eventual disappearance of Anasazi culture, for archaeologists have ascertained that the great houses at Chaco, including Pueblo Bonito, had been abandoned by the start of the thirteenth century—seventy-five years before the severe late thirteenth-century drought set in. But unreliable rains may have combined with other forces in undermining this culture. The Chacoan system probably collapsed under its own weight. Evidence suggests that Anasazi populations increased dramatically in the eleventh century, and these increases may well have outpaced food production. Some archaeologists believe that Chaco not only was a political and ceremonial capital but also served as an administrative and distribution center for the region's food supply: 400 miles of roads linked Chaco Canyon with outlying satellite pueblos. The Chacoan political and social structures may have been unable to cope with so vast and complex a system, especially in dry years. Overcultivation depleted soils in favored alluvial valleys and made the corn fields vulnerable to erosion. Even in a good year, there was simply not enough arable land or water to raise sufficient food for all the people. As early as the first decades of the twelfth century, Chaco, with its great storerooms and immense conical underground ceremonial chambers known as kivas, presided over a hollow empire. By the middle of that century, the empire had been dismantled.

Architectural historian Vincent Scully speculates that the increasing size and openness of the kivas in the twelfth and thirteenth centuries and their placement at watercourses may have been a kind of urgent architectural invocation of the rain deities—louder and grander prayers to bring rainfall commensurate with the bigger and wider kivas. "[I]t is likely," writes Scully, "that progressive desiccation, culminating in the decisive drought of the later thirteenth century, led to intensified attempts to draw yet more of the earth and the sky, as well as more of the larger human community, into the web of ritual."[13] It's possible that when prayers for rain failed for several seasons running, the Anasazi concluded that the water spirits had forsaken their pueblo and so they abandoned it. Or perhaps they stopped making pilgrimages to Chaco because they believed the priests had lost touch with the elements. As Kendrick Frazier points out, Chaco managed to dominate the region in part because the people believed that precious rainfall sprang from the yearly rituals performed there: success of the rituals "would reinforce the perceived value and power of the system," according to Frazier, while their failure would undermine it.[14]

Drought was the Achilles' heel of Chaco Canyon. When the rains ceased to fall, the system centered on the great pueblos expired. Though the Anasazi had invested tremendous manpower and artistry in constructing the pueblos, they may have felt that it was futile, indeed blasphemous, to continue living there without supernatural sanction, and so they moved on to new sites that they hoped would be more favorable. "Nature is our home," as one contemporary Native American told me. "If it becomes too hostile we shouldn't be living there. The Anasazi would have left if their pueblos became too unproductive."

There is considerable evidence of such departures starting about 1150, when the Anasazi began regrouping in fewer but larger pueblos to the south and east of the Four Corners region where their culture had been centered. This pattern accelerated during the drought years of the late thirteenth century. As Alfonso Ortiz writes, "During the persistent drought that haunted the Southwest in the last quarter of the thirteenth century, the Pueblo people began contracting into the great valley of the Rio Grande and its tributaries. Only the Hopi, then as now, enigmatically hung on and persisted in farming successfully in a region with no permanent or semipermanent watercourses."[15] Another researcher postulates that the Anasazi banded together in larger communities at this time in order to mount ever more elaborate rain rituals: "The adaptive difficulties that they were experiencing would have made Pueblo groups receptive to adopting or developing new ceremonies in search of supernatural assistance."[16] The cult of Tlaloc, the bloody Aztec water deity who exacted child sacrifices in time of drought, began to take root in the American Southwest about this time—another sign of the Anasazi's desperation to influence rainfall. But supernatural assistance failed to materialize, and the culture continued to decline. When the Spanish arrived in the Southwest in the sixteenth century, the Anasazi pueblos were empty and the peoples had dispersed. Even their name was lost: Anasazi is a Navaho word translated variously as "the Old Ones," "the ancient ones," or "someone's ancestors," a name invented by migrants from the north for the mysterious people who had lived on this land before them.

I believe I have grounds to say that the known history of American weather begins with the Anasazi. Of climate history, we know something —and in some cases quite a lot—going back to the Ice Age and even before. But of *weather* in the sense of man's subjective experience of climate, our first tangible evidence surfaces in the ceremonies, prayers, gods, migration patterns, and perhaps the kiva design of the Anasazi. Prayer, ritual, cosmology, architecture, town planning: what are these but efforts to interpret, influence, explain, and adapt to climate? Would it be too farfetched to suggest that the increasingly elaborate ceremonialism of the last generations of Anasazi is a forerunner of our own ever more elaborate and technologically sophisticated meteorology—our bold, obsessive attempt to predict the unfathomably chaotic? Perhaps. After all, the evidence is still too sketchy, our insights too speculative. We know that the Chacoans were advanced enough to align the apertures on the south wall of Pueblo Bonito precisely with the path of the sun on the summer solstice. The "sun dagger" recently discovered on the top of Fajada Butte was an even more precise tool for charting the yearly cycle of the sun's path. The shape and the placement of stone tell us this much. But the ruined stone cities keep their secrets about the beliefs, attitudes, emotions, and convictions of the people who built them. Our first window into the history of America's weather, if indeed it is a window, is small and cloudy.

A few more rays of clarity penetrate when we compare our guesses about Anasazi weather beliefs with the rituals of their descendants, the so-called Pueblo Indians who still live in the Southwest—the Hopis and Zunis of Arizona and New Mexico and the peoples who have lived for centuries in the pueblos of Acoma, Santa Clara, Laguna, Santo Domingo, San Juan, Taos, Picuris, and other sites. Ceremonies associated with rainfall and water remain central to the culture of these people—indeed, as Hartley Burr Alexander writes, "in many ways the entire culture seems built up out of a prayer for rain."[17] Perhaps the most famous, and certainly the most extensively documented, of these prayers for rain is the Hopi snake dance performed each August at the start of the rainy season. In the ritual, male members of the snake and antelope fraternities disperse in the four directions to gather as many snakes as they can find. The snakes are placed in a bower of cottonwood limbs in the pueblo's central plaza, and at the climax, the snake priests pick up the snakes one by one, place them in their mouths, and dance with them. The priests repeat this until they have danced with every snake—in a good year as many as seventy or eighty. Supposedly, the Hopis chose snakes as the central symbol in this ritual because they associate rattlesnakes with the jagged pattern of lightning bolts—and from there the association spread to rain and snow, rivers and springs, tree sap and human blood. Rattlesnakes are thus the snake of choice for the snake dance, but bull and garter snakes will also do. In the dance, snakes are believed to be transformed into messengers to the gods of the underworld: they implore the supernatural powers for rain and report that the Hopi celebrants are worthy of receiving rain for their crops because they have remained true to the ways of their ancestors. The Hopis and other Pueblo Indians view such rituals not merely as prayers but as an essential element in maintaining the vital equilibrium of the world. When thousands of Pueblo Indians died during a severe drought that lasted from 1663 to 1669, the Indians blamed the Spanish for interrupting their rituals and thus destroying the ancient harmony between man and the forces of nature. Cosmic cataclysm was the inevitable result.[18]

When the Hopis pray, they include in their prayers not only all Hopis, living and dead, but all of humanity. Weather, and especially sufficient summer rain, is perhaps the most important manifestation of the cosmic order to which all beings, living and dead, animate and inanimate, belong. Ritual is our role in this order; rainfall is the visible manifestation of the supernatural powers: but ritual and rain are linked not so much by cause and effect as by their common place on the same continuous scale. The Hopis dance with snakes in their mouths as much *because* it rains as to make it rain again. As Ake Hultkrantz, an authority on Native American religions, writes, rain rituals are not only a way of fostering a good harvest but "a way of taking part in the drama of the emergence and reemergence of plant and animal life."[19] "We don't praise the Ultimate," one Native American told me. "We *recognize* the Ultimate." Weather is thus not separate from and outside of man—something that happens *to* us—but one of the guises that our participation in nature assumes.

The association of rain with blood and with the spiritual essence of living things also shows up in Hopi mythology and worldview. The Hopis believe that the spirits of the dead return as kachinas—cloud people whose spiritual essence, their "navola," is expressed as rainfall. "Everything, in Hopi belief, is dependent on rainfall, which, when combined with Mother Earth, is the essence of all things," writes one scholar of Hopi ways. "Hence navola is also the essence of the individual self, conceived of as a liquid, and a Hopi will say, 'I have the liquid essence of my fathers' to express the English notion of being of the same flesh and blood. Through the combination of the rain with the earth and its transformation into corn, the blessings of the kachinas (their navola) become the essence of our bodies (our navola). There is, thus, an essential consubstantiality in the bipartite structure of the Hopi universe that relates cotton masks and clouds, the living and the dead, rain and life."[20]

The Hopis, living as they do on a series of extremely dry and desolate mesas, have made rainfall rituals almost a cult and have made liquid the prevailing metaphor of their mythology. But weather ceremonialism and water imagery are widespread in the rituals and myths of all southwestern peoples—and often linked to fertility rites. In the extremely elaborate Zuni religion, members of sixteen rain priesthoods dedicate themselves from July to September to the encouragement of rainfall through quiet meditation. The Zuni kachinas, unlike the kachinas of the Hopi, are spirits of the dead who play a critical role in bringing rain and good harvests. Hultkrantz notes that the Zuni kachina cult is in fact a form of ancestor worship: by making their dead ancestors rain gods, the Zunis knit their lives into a fabric that stretches through time.[21] The famous corn dance that members of the Santo Domingo pueblo southwest of Santa Fe perform each August 4 is a rainfall ritual in which dancers wear costumes decorated with clouds, raindrops, lightning, and serpents. "Now come rain! Now come rain!" chant the celebrants. "Fall upon the mountains; sink into the ground." "Come snow in plenty," the villagers at San Ildefonso pueblo sing each winter, "that water / may be abundant when summer comes." There are many other examples of southwestern weather rituals with roots reaching way back in time. Maybe all the way back to the Old Ones. Kendrick Frazier is convinced that the weather rituals that the Pueblo

Indians practice today are directly descended from weather rituals prac-
ticed in Chaco Canyon and other Anasazi sites 1,000 years ago: "Pueblo
ritual and ceremony—and I feel certain this was true of their ancient
ancestors as well—have at their heart the necessity to live in harmony with
the world."[22]

NATIVE AMERICAN PEOPLES IN OTHER SECTIONS OF THE COUN-
try also revered and beseeched weather phenomena, but outside the arid
agricultural Southwest, weather ceremonialism has played a less central
and less dramatic role in the culture. Indians of the East Coast and the
prairies occasionally performed rainmaking rites, but usually these were
the work of a single shaman rather than part of a communal celebration.
Where rainfall was not critical for the survival of the tribe, rainmaking
could be a feat of magic that the shaman used to enhance his personal
power and prestige.[23] In historic times, skeptical whites have witnessed
such feats for themselves. To cite but one dramatic example, the Yurok
tribe of northern California have a rain stone that was so potent in bringing down precipitation that they buried it; when a road crew unearthed it
in 1959, five inches of water poured down the next day. Floods followed a
subsequent digging up of the stone in 1966.[24] Of course, rain sometimes
fails to follow a rain dance or the striking of a rain stone—but, as Native
American rainmakers point out, rain has also been known to fail to arrive
after the National Weather Service issued a "100 percent chance of precipitation" forecast. As far as I know, there have been no studies to date
comparing the accuracy rates of Indian rituals and NWS computer-generated forecasts.

Indian mythologies, folk tales, and religious symbols hold other,
fainter clues about weather attitudes and beliefs. The thunderbird is a
potent spirit in many Indian tribes, particularly in the Plains states, where
summer thunderstorms are among the fiercest on the planet. A great
winged creature often depicted as an eagle, the thunderbird has dominion
over thunder, dark clouds, rain, and snow. Thunder resounds from the
flapping of his enormous wings, lightning flashes from the winking of his

eyes. He is a kind of Indian Zeus, and his yearly arrival in the first thunderstorm of spring is an occasion of great import. The Wind River Shoshoni of Wyoming believe that the thunderbird controls some winds, while other winds blow at the will of the *hyipij,* the wind master who lives in the highest mountains.

The Iroquois of the Northeast assign an animal to each of the four winds—Bear to the fierce winds that blow from the north; Panther to the snarling blasts from the west; Moose to the damp, misty east wind; and Fawn to the soft south wind that heralds the spring. The Yuki Indians of California believe that the supreme being makes himself visible through flashes of lightning and speaks through thunder; the rainbow is his clothing.

In the Pacific Northwest, Indians used to blow whistles at the start of the damp, chill winter season to signal the approach of supernatural beings: while spirits floated with the mists through the immense conifers that pressed closed around the villages, the Indians devoted themselves to wood carving, painting, ritual observances, and spirit quests. Pacific Northwest mythology explains the region's soggy climate as the result of a bargain struck between the animals and the petulant South Wind: the animals bit and clubbed the South Wind in his home after he blew without cease for weeks, and to save his life, the South Wind agreed to alternate fine weather with nasty—four sunny days would follow four dismal ones. According to the Hoh and Quileute peoples of the Olympic Peninsula, the prism of colors that arcs through the sky after a rainstorm is a manifestation of the Rainbow Woman, who climbs the sky from beyond the ocean to look for her lost human lover. Many Pacific Northwest peoples have variations on a Great Flood myth, perhaps a dim racial memory of the rising of the ocean at the end of the Ice Age.

The Ojibwa or Chippewa of the northern Great Lakes also have a myth that could refer back to postglacial climate change: they tell of a time when the world was so cold that snow buried the spruce trees and the people followed Squirrel into the sky on a quest for warmth; their quest led them to a bunch of leather bags containing rain, snow, storms, cold, and sunshine. Seizing the bags, they headed back home, but Squirrel tore up one of the bags to patch his moccasin. Warmth rushed out of the bag and quickly melted the snow, and the entire earth was flooded. It was Duck who rescued mankind by bringing up mud from the bottom of the waters and creating new lands.[25]

It's curious how powerful animals are in the weather myths of the American Indians. Unlike the Old Testament story of Noah, in which a man acts as God's agent in saving the animal kingdom from disastrous weather, the Indian tales give the creatures control and put mankind at their mercy. Standing one step closer to the elements, the animals influence and embody weather phenomena; through their spirits, we are joined as kin to the thunder, the snow, and the hesitant, life-giving winds of spring.

Most Native peoples living in America today have access to National Weather Service forecasts and warnings. They push a button, like the rest of us, to hear long-range forecasts and winter storm advisories. Yet the weather beliefs and attitudes of their ancestors remain alive. Spirits have not vacated the clouds and winds of our continent. Rain still falls from dancing. The sun is still prayed up into the sky each morning. Ed McGaa, an Oglala Sioux Indian and the author of *Rainbow Tribe* and *Mother Earth Spirituality,* described the persistence of the old beliefs among his people:

"The four directions is a concept that is generally accepted among Indians. Different types of weather reside in each of the four directions. The North is a giant, and his color is white because of the snow. The West brings rain. The South is the direction of the sun—one watches it get higher and higher on the horizon after the winter solstice. The East brings knowledge. Thunder and lightning we see as a power that comes out of the West. We say the West power has the 'thunder being' in it. The West has the power to make life and destroy it. The North makes the earth rest. The South makes it grow.

"The Indians believe that the sacred goes inside of science—that the sacred is another truth—unlike the white man who tries to make physics and chemistry the only truth. For Indians, weather is a part of the whole cycle of nature—it's a means that the creator has devised for us to live. Weather is controllable to some extent. Nature becomes more responsive to us the more we know of it. We have influence over the weather—we have a track record to prove this, certainly a more successful track record than many others. At the sun dance, Indians call for long cloud and the cloud appears. I don't claim to understand it, but it happens. The elements acknowledge and communicate with man. The nearest I can come to explaining it is to say that all being is responsive to the ultimate creator, and the creator allows humans to influence the four major forces through spiritual connection. It's like with the animals—the more we know them, the more they trust us and gather around us. Weather is just a degree removed from the animals. Weatherwise, the Indians beseech the four directions for the type of weather they seek. This is especially true for crop-raising Indians."

Some dismiss contemporary Indian rainmaking rites as showmanship, claiming that they schedule the ceremonies when they know a storm is blowing in. Perhaps another way of looking at it is to say that Indians have learned to be extraordinarily accurate weather forecasters because of their ancient tradition of attunement to nature—a tradition born of necessity and sustained out of reverence for the ancestors. To take but one well-known example: Back in the 1920s, Chief Billy Bowlegs of the Florida Seminoles became famous for his ability to predict hurricanes. Chief Bowlegs led his people to shelter deep in the Everglades in September of 1926 just days before a hurricane swept across the Florida coast, killing 372 people and destroying the Seminole encampment at the Loxahatchee River. (Zora Neale Hurston dramatized an even more severe 1928 storm in her novel *Their Eyes Were Watching God*: "The monstropolous beast had left his bed. The two hundred miles an hour wind had loosed his chains. He seized hold of his dikes and ran forward until he met the quarters; uprooted them like grass . . . rolling the dikes, rolling the houses, rolling the people in the houses along with other timbers."[26]) Later, Chief Bowlegs identified some of the signs that tipped him off to the approach of the storm: an unseasonable blooming of saw grass; the migration of rats and rabbits north and west of the coast; an unusual silence among the birds and abrupt flights northwest; an eerie barking of local alligators and movement into deep water. An official of the Weather Bureau at Miami commented,

"Sometimes our predictions are inaccurate because it is impossible to forecast the capricious turns these storms often take. But those Indians never miss."

"Nature becomes more responsive to us the more we know of it," says McGaa. For thousands of years, this central belief determined the prevailing weather in America. And then European explorers crossed the Atlantic Ocean, and America's weather changed.

2.

WEATHER IN THE AGE OF DISCOVERY

ABOUT A MILLENNIUM AGO, WHILE ANASAZI INDIANS WERE quarrying sandstone blocks from the walls of Chaco Canyon and Mississippians were creating enormous ceremonial mounds at their stronghold at Cahokia in southern Illinois, the Europeans basked. From about 800 A.D. to 1200 A.D., the ordinarily cool, damp climates of Britain, northern France, Germany, and the Netherlands turned Iberian. Summers were sunny and warm, and sometimes torrid. Winters were mild. Drought was common. During the centuries of the Medieval Optimum, as this period is known, average temperatures in the Northern Hemisphere may have been as much as 4° F higher than today—higher than they ever got during the headline-making hot years of the late 1980s. The medieval warmth could not be blamed on cars and factories belching greenhouse gases into the atmosphere or on widespread slashing and burning of tropical rain forests: it just happened. Odd that what we deem optimal for the "Dark Ages" we consider a portent of disaster for ourselves—but such are the ways of science.

The climate change of the Medieval Optimum was significant enough to alter the surface of the land and the sea. In England vineyards flourished as far north as York, and the wine they produced was judged the equal of what came out of France. Glaciers retreated in the Alps and trees took root on mountain slopes that a century before had been buried under year-round ice. Ice broke up on the northern seas, opening new passages to navigation. Norse mariners, the most skilled and adventurous sailors of the day, promptly seized the opportunity offered by nature. The image that the word "Norse" brings to mind is a blond pirate in horned-helmet and fur robes, but historians report that this is false: the Norse who were moving north and west in the ninth and tenth centuries were actually peaceful farmers and fishermen, not raiders. They reached Iceland around 860 and planted colonies there soon after. In the final decades of the tenth century, Norse parties led by Snaebjorn Galti and Eric the Red crossed due west from Iceland to the vast arctic island they named Greenland (exaggerating the verdure with an eye to luring colonists). Here too they established colonies and plowed up farms. And then, inevitably, the Norsemen breached the fastness of the next landmass west—the northeastern fringes of North America. Millennia of Ice Age cold opened a bridge over which the Paleo-Indians walked to America; four centuries of medieval warmth drew Europeans to the New World by sail. Climate change was crucial to the second "discovery" of America, as it had been to the first.[1]

Leif Ericsson, Eric the Red's son, set sail for the mysterious west in the year 1001. Most historians, comparing the Icelandic sagas with the evidence of geography and archaeology, believe that Leif first sighted land at Baffin Island and later went ashore at Labrador. The sagas record that the party then proceeded west until they came to another spot that pleased them much. The land had an open meadow where they could graze their cattle, a stream running with large and plentiful salmon, and, surprisingly, an abundance of grapes, which gave Leif a name for the country—Vinland. For years historians debated the location—and the very existence—of this mystical land of wine. Wine grapes would not have flourished north of

Cape Cod, even during a prolonged mild period, so some historians concluded that Leif Ericsson must have sailed quite far south, perhaps all the way down to the Chesapeake. Others chalked Vinland up to the vivid imagination of the anonymous saga writer who recorded Leif's exploits two centuries after they had occurred. But then, in the 1960s, archaeologists made an almost too-good-to-be-true discovery that settled the matter. At L'Anse aux Meadows on Epaves Bay at the northeastern tip of Newfoundland an archaeological team uncovered remains of a Norse settlement that bore an unmistakable resemblance to the Vinland of the sagas. The great houses with central fire pits, the meadow, the salmon stream—it was all just as the sagas described it. The one large question that remained unanswered concerned the grapes: the meadow at L'Anse aux Meadows grows cranberry, partridgeberry, cloudberry, bearberry, and bunchberry, but no grapes. Samuel Eliot Morison speculates that the "vin" of "Vinland" really referred to red currants or gooseberries; others say that word has been mistranslated and actually means grassland, nor wineland.

In any case, the finds at L'Anse aux Meadows established the veracity of Vinland and led scholars to the conclusion that the sagas were more truthful than had long been supposed. The Norsemen crossed the northern sea, planted a North American colony, fished, explored, and encountered Indians (whom they called Skraelings, which probably meant barbarian or pygmy), just as "The Saga of Eric the Red" in "Hauk's Book" reported they did. And, inevitably, Leif Ericsson and his followers sized up the weather of this strange new world. "The Greenlander's Saga" (*Groenlendinga Saga*) gives the first tantalizing glimpse of American weather through the eyes of Europeans: "Nature was so generous here that it seemed to them no cattle would need any winter fodder, but could graze outdoors. There was no frost in winter, and the grass hardly withered. The days and nights were more nearly equal than in Greenland or Iceland. On the shortest day of winter the sun was up between breakfast time and late afternoon."[2] It is telling indeed that the New World should make its first appearance in literature as a kind of Eden—a land blessed with natural abundance, ever-green meadows, bright, long days, and mild weather. The belief, born of hope, that America was a paradise on earth, complete with paradisal weather, would persist for hundreds of years, indeed into today—coexisting with the image of America as a howling, violently tempestuous wilderness.

Vinland proved to be a very brief experiment. Other Norse groups followed Ericsson, but nature's generosity was not sufficient to pull the colonists through more than a couple of winters; food supplies grew scarce; infighting thinned the Norse ranks; and the Skraelings, at first curious about the white-skins and eager to trade furs for red cloth, turned violent. A huge Skraeling war party paddled up from the south and attacked. Many Norsemen perished under a "fierce shower of missiles" launched by the Skraelings' "war slings." As the saga reports, "It now seemed clear to Karlsefni [a connection of Leif's and the leader of a subsequent Norse mission to Vinland] and his people that, although the country thereabouts was attractive, their life would be one of constant dread and turmoil by reason of the inhabitants of the country. So they forthwith prepared to leave, and determined to return to their own country." And so ended the first documented and substantiated European colony in North America.

It is possible that the Norse might have tried to colonize Vinland again if their fortunes held in Europe. But by the late twelfth century, circumstance—and most of all the weather—had turned against them. The Medieval Optimum that opened Iceland and Greenland to Norse settlements ended in a series of damp and increasingly chilly years. After 1160, drift ice once again began to block the passage of westbound ships. Northern harbors iced over. The Norse colony in Greenland gradually, and agonizingly, starved and froze to death during the twelfth and thirteenth centuries. Meteorologist Reid A. Bryson in his book *Climates of Hunger* attributes the death of the Norse colony in Greenland to the same global weather pattern that drove the agricultural Indians off America's Great Plains. Bryson blames these phenomena on a prolonged shift in the westerlies—the upper-atmosphere current of air that flows west with the jet stream around the North Pole.[3] Typically, the westerlies contract toward the pole during the summer and expand in great loops to the south during the winter. But starting in the twelfth century, the westerlies remained in a slightly expanded winter-type position during the summer months, allowing arctic air to maintain its grip over Greenland, drawing cool moisture-laden ocean breezes into northern Europe, and preventing humidity from the Gulf of Mexico from blowing north onto the Plains. As a result, England, Germany, and France experienced a series of very poor harvests with too much rain and too little sun to ripen the grain; the American Plains and Southwest had drought; drift ice closed the harbors of Iceland and Greenland and cold weather made farming increasingly difficult. This pattern remained in place for a century or so, and may have caused the drought that broke the back of Anasazi culture as well. By the time the westerlies resumed their usual pattern, the Indian villages of the central Plains had broken up, the Anasazi had dispersed south and west, Europe had endured serious food shortages and widespread spoilage of grain, and the last few stunted Norsemen in Greenland fell prey to disease, hunger, and warfare with the natives.

The medieval century of climate change was nowhere near as severe as the great swing to warmth that ended the Ice Age; but it was marked enough to cause upheavals in the lives of people inhabiting marginal regions—the arid Plains, the desert Southwest, the subarctic north. In the centuries to come, the weather would shift again, and even more drastically, over the Northern Hemisphere. And once again, the shift would coincide with—and perhaps precipitate—a push to explore and colonize the New World.

THE WINTER OF 1607–8 WAS A SEASON THAT CARVED A PERMANENT place for itself in English history. The Thames River froze to such a depth and stayed frozen for so long that the people of London staged a spontaneous hibernal revel on the back of the paralyzed river—the first Frost Fair. The playwright Thomas Dekker wrote a play inspired by the season called *The Great Frost*. Shakespeare's tragedy *Coriolanus* contains the line "the coal of fire upon the ice," thought to be an allusion to the pans of coal burned on the river ice that winter. "The Great Frost was, historians tell us, the most severe that has ever visited these islands," writes Virginia

Woolf in her novel *Orlando*. "Birds froze in mid-air and fell like stones to the ground. At Norwich a young countrywoman started to cross the road in her usual robust health and was seen by onlookers to turn visibly to powder and be blown in a puff of dust over the roofs as the icy blast struck her at the street corner. . . . Corpses froze and could not be drawn from the sheets. It was no uncommon sight to come upon a whole herd of swine frozen immovable upon the road."

This is fanciful, of course, but not entirely fantastic, for it does reflect the way climate change enters the popular imagination. The Great Frost, we now know, was not an isolated occurrence, but rather a sign of a marked and prolonged shift downward in the average temperature of Northern Europe. For generation after generation during the medieval period ice was never known to stop river traffic on the Thames. But then, rather suddenly, starting in the 1560s, the Thames froze over for part of many winters, often freezing and thawing several times during a single winter. These freezes continued until February 1814, and have not re-sumed since. The Little Ice Age, as this chilly interval has come to be known, brought the Northern Hemisphere its lowest average temperatures since the retreat of the glaciers—as much as 5.4° F colder than our average temperatures in the coldest years. "Historians of climate differ in their dating of the Little Ice Age," notes Nigel Calder in his book *The Weather Machine*, but most would agree that the worst of the cold came roughly between 1430 and 1850.[4] After the Great Frost of 1607–8, Frost Fairs became a London tradition, culminating in the great fair of 1683–84, a notoriously rigorous winter during which the ice froze ten inches thick and remained intact for two solid months. While Londoners were dining and cavorting on the frozen Thames, the Dutch skated on their iced-over canals and the French were reduced to chipping their wine out of casks and distributing it in hats. Up in the Alps, glaciers advanced down the mountainsides, expanding even in the summer during particularly cold years. Villages were swallowed, crops destroyed. "We have also heard it said that there are evil spells at work among the said glaciers," reported a French official of the advancing ice front on Mont Blanc in the spring of 1642, "and that the people, last Rogation-tide, went in procession to implore God's help to preserve and guarantee them against the said peril."[5] Several hundred years earlier, the Indians of the American Southwest had gone in processions of their own to implore their god's help against the peril of drought. The parallels would, no doubt, have been lost on the Europeans.

Frost fairs and glacial advances are historically documented in journals, literary works, and paintings of the period: we don't need tree rings or pollen fossils to tell us the world was colder for the better part of four centuries. Yet many climatologists, while granting the chill, object to the term Little Ice Age. The cold of these years was neither as universal nor as relentless as the name implies—and conditions were certainly tepid compared to the glacial advances of the Pleistocene. The seventeenth century had its share of torrid summers—the most famous being the long, hot, dry summer of 1666 when a desiccated London went up in flames. And mild winters did occasionally break the pattern of frigidity. The Pilgrims were blessed with an exceptionally mild winter after their landing at Plymouth in 1620—"a calm winter, such as was never seen here since," as one of the colonists described it; but even so half of the original settlers were dead before summer. The diarist Samuel Pepys commented with some astonishment on the extraordinary clemency of the winter of 1660–61, when England remained frost-free and roses were in full leaf in January. And yet the cold of the Little Ice Age was, for long stretches, intense enough and *different* enough from what had gone before that people took notice. Something had "happened" to the weather. This was not a slow, gradual cooling trend, but a rather abrupt and extended plunge into markedly colder weather. As we learn more about past climates, we are uncovering evidence of other instances of these sudden swings. It is not only the earth's weather but also its climate that changes constantly and erratically. What is remarkable about the Little Ice Age is not that it happened, but that it happened around the same time that Western Europeans were exploring the world. It is my contention that these natural and human events were closely linked.

• • •

THE LITTLE ICE AGE, THOUGH NOT AS SEVERE AS THE NAME suggests, got people thinking about the weather. It set them at odds with an environment that seemed to have betrayed them. When frigid winters were followed for several years running by damp, cool summers with poor harvests, that anxiety turned to disgust, alarm, even panic. "Intolerable severe frost" English diarist John Evelyn grumbled in the 1680s. "Excessive cold . . . backward spring . . ." It wasn't *supposed* to be that cold in England. The cold spawned nostalgia for the "good old days" and increased demand for the exotic produce of tropical and subtropical regions. One catalyst for global exploration was the desire to open easier, cheaper routes for the importation of goods that had grown increasingly valuable during the cold centuries—silk, spices, gems, and precious metals; and later, in the seventeenth century, European traders penetrated deeper and deeper into the North American interior in pursuit of beaver pelts, which were used to make beaver-felt hats, a highly prized and useful status symbol in a cold climate. The poor harvests of the Little Ice Age sent shock waves through the economies of Europe. The people had become restless on their cramped, cloudy, snowbound continent. The edgy music of change was in the air. New trade routes. New science. New wealth. New World. Could it have been that the meteorological swings of the Little Ice Age drove European mariners literally to the ends of the earth? It's certainly an intriguing possibility.

In any case, whether or not the Little Ice Age helped propel them, the explorers came in a trickle and then a torrent after Columbus crossed the Atlantic in 1492: John Cabot, an Italian sailing for England, touched the coast of Newfoundland, possibly a few miles away from Leif Ericsson's Vinland, in 1497, and continued south, perhaps as far as New England; a series of Portuguese navigators made the transatlantic trip around the start of the sixteenth century; Sebastian Cabot followed in his father's wake about 1508; and in 1524, Florentine navigator Giovanni da Verrazano, sailing under the flag of the French King Francis I, made the first European voyage up the Atlantic coast of what later became the United States between North Carolina and Maine. Jacques Cartier, John Hawkins, Sir Francis Drake, Sir Humphrey Gilbert, Bartholomew Gosnold, Martin

Pring, and Samuel de Champlain followed in the course of the late six-teenth and early seventeenth centuries.

❡   "Psychologists tell us that we see what we are expecting to see," writes Evelyn Stefansson Nef in the introduction to a collection of docu-ments written by the early European explorers. "When one of our explorers sees a uniped, cyclops or a mermaid . . . you may be sure he was expect-ing to see one."[6] The early explorers certainly saw the weather they were expecting to see in the New World. Since they believed they had reached the fringes of the Orient, they naturally perceived something Oriental about the weather that greeted them on the coast of America. "This land is in latitude 34 d[egrees], with good and wholesome air, temperate between hot and cold," Verrazano wrote to King Francis I in the early summer of 1524 of his impression of North Carolina's Outer Banks, which he be-lieved to be the only land that separated the Atlantic from the Pacific Ocean; "no vehement winds do blow in those regions, and those that do commonly reign in those coasts are the northwest and west winds in the summer season (in the beginning whereof we were there), the sky clear and fair with very little rain. And if at any time the air be cloudy and misty with the southern wind, immediately it is dissolved, and waxeth clear and fair again. The sea is calm, not boisterous, the waves gentle."[7] Notorious for severe storms, gale-force winds, crashing surf, and treacherous currents that have devoured many a ship, the Outer Banks were clearly enjoying a tranquil interval during Verrazano's sojourn there. It's likely, too, that he was embellishing the softness and mildness of the weather a bit to entice the king's interest, for he also reported that the tawny color of the beach sand promised rich deposits of gold nearby. One interesting detail of Ver-razano's report that many later explorers confirmed was the delightful smell that wafted out to sea from the shores of North America. "We smelt so sweet and so strong a smel, as if we had bene in the midst of some delicate garden abounding in all kinde of odoriferous flowers," wrote the English captains Philip Amadas and Arthur Barlowe in 1584 of their approach to Roanoke Island, the site of the "Lost Colony" sponsored by Sir Walter Raleigh.[8]

Verrazano did admit that the weather got colder as his party cruised north up the coast, and he identified the trees growing along Maine's shore as "suchlike as are wont to grow in cold countries." Verrazano's descriptions of America's topography, natives, and climate are as vivid and richly colored as Mannerist painting; but one senses at the back of them a strong desire to please his royal patron—and to justify the considerable expense of the voyage. The weather *had* to be "as pleasant as is possible to declare" and the land brimming with "drugs . . . spicery and other riches of gold"—otherwise why bother venturing halfway around the world?

Jacques Cartier, a Frenchman who followed Verrazano with voyages of exploration in 1534, 1535–36, and 1541–42, was the first European to experience and record the truth about a North American winter. On his second voyage, Cartier and his men wintered over at Ste. Croix near present-day Quebec, building a stockade and laying in what supplies they could. In his record of this winter, Cartier describes how a combination of severe cold and scurvy decimated his company: "From the midst of November until the midst of March we were kept in amidst the ice above two fathoms thick and snow above four feet high and more, higher than the sides of our ships, which lasted till that time in such sort that all our drinks were frozen in the vessels, and the ice through all the ships was above a hand-breadth thick, as well above hatches as beneath; and so much of the river as was fresh, even to Hochelaga [Montreal], was frozen; in which space there died 25 of our best and chiefest men, and all the rest were so sick that we thought they should never recover again."[9] The Indians saved many of the French from death by scurvy by teaching them how to make a remedy out of the boiled bark of the white cedar tree; but they could not teach the French to endure the rigors of winter. Cartier's party was astonished to see how comfortable and healthy the Indians remained through the frigid months, despite the fact that they went out in ice and snow almost entirely naked.[10] The cold proved to be a formidable enemy to all of the initial European attempts at colonization.

"It was difficult to know this country without having wintered there," wrote Samuel de Champlain of the winter of 1604–5 on St. Croix island between Maine and New Brunswick; "for on arriving in summer everything is very pleasant on account of the woods, the beautiful land-

Lightning strike, March 6, 1969. About one hundred people die in lightning strikes every year in the United States, making lightning a greater killer than tornadoes or hurricanes. (NOAA)

A tornado demolishing a farmhouse on the Plains. An average of seven hundred tornadoes strike the United States each year, the majority of them ripping through the "tornado alley" that runs from north-central Texas through Oklahoma, Missouri, Kansas, Nebraska, and north into the Dakotas. (NOAA/NWS)

Hurricane storm wave near Miami Beach, Florida. *(NOAA)*

The steeple of Boston's Old North Church (the church in which Paul Revere hung the warning lantern when the British were coming) succumbing to winds generated by Hurricane Carol on August 31, 1954. An 1804 hurricane had knocked off the original steeple, which dated back to 1723. *(Library of Congress)*

Cumulonimbus clouds—also known as thunderclouds—
often bring heavy rains and high winds. *(United States Weather Bureau)*

Native peoples have been performing weather rituals in America for millennia. In this 1903 photo, Hopi jesters and rainmakers celebrate the onset of the summer rainy season at the pueblo in Oraibi, Arizona. *(Library of Congress)*

Chief Turtle dancing for rain at the Indian celebration in the Two Medicine Valley of Glacier National Park. Torrential downpours arrived several hours after this picture was snapped. *(Bain Collection, Library of Congress)*

An early twentieth-century photo of the Hopi snake dance, one of the most famous and extensively documented Native American rain rituals. At the climax of the ceremony, the snake priests place the snakes one by one in their mouths and dance with them. The Hopis believe that in the course of the ritual the snakes are transformed into messengers who descend to the underworld and implore the gods to bring rain. *(Library of Congress)*

Increase Mather, staunch Puritan, father of staunch Puritan Cotton Mather, and pastor of Boston's North Church for sixty-two years, was obsessed with extreme weather. Fully three chapters of his 1684 volume *Remarkable Providences* deal with remarkable weather. *(Library of Congress)*

Ben Franklin and son flying the most famous kite in American history, June 1752. Franklin's experiment "compleatly demonstrated . . . the Sameness of the Electric Matter with that of Lightning," and thereby revolutionized our understanding of weather. *(Library of Congress)*

Thomas Jefferson, scientist. A passionate amateur meteorologist and an eloquent defender of America's climate, Jefferson kept detailed and meticulous weather records throughout his adult life. He was convinced that science could "ascertain the climate of the United States and its probable causes" if only sufficient data was amassed and analyzed. *(Library of Congress)*

The enigmatic explorer Zebulon Pike, who may have been a spy, made a vast circuit of the American Southwest in 1807. He described the region as the "Great American Desert," a term that frightened off settlers for half a century. *(Library of Congress)*

John Wesley Powell and his exploring party about to push off down the Green River in 1871. Powell recorded his conclusions about the topography and climate of the West in his authoritative *Report on the Lands of the Arid Region of the United States,* the book that broke the news to Americans that much of the West was too dry to farm without irrigation. *(Library of Congress)*

Settlers on the Great Plains confronted a weather phenomenon largely unknown in the East— the twister. *(NOAA/NWS)*

A South Dakota farm after the dust storms of 1935. More than four and a half tons of topsoil *per acre* entered the atmosphere during the five black blizzards that swept through the Plains that spring. *(NOAA)*

Dust bowl, Cimarron County, Oklahoma. When asked to comment on this photo as part of a sociological study conducted during the 1960s, one farmer replied: "The man [is] going to tell those boys, 'Well, we didn't raise any crops this year but wait till next year.' Because that is the biggest next-year country in the world out there."
(Photo by Arthur Rothstein, Library of Congress)

Reporter Robert Geiger coined the term *dust bowl* in a story he wrote after the notorious Black Sunday of April 14, 1935: "Three little words, achingly familiar on a western farmer's tongue, rule life today in the dust bowl of the continent—if it rains." *(NOAA)*

Black blizzard about to engulf Elkhart, Kansas, May 1937. *(Library of Congress)*

Undated dust storm. *(NOAA)*

scapes, and the fine fishing. . . . There are six months of winter in that country. . . . The cold was severe and more extreme than in France, and lasted much longer."[11] The "dreadful winter" cold forced the French to abandon their 1604 colony and reestablish it on the more temperate (though hardly balmy) site of Annapolis, Nova Scotia. Winter cold also evicted one of the first groups of English colonists. In August 1607, 124 English settlers under the leadership of George Popham established a colony on the banks of the Kennebec River on Maine's coast, not far from the town of Phippsburg. Had it survived, the Popham colony would have gone down in history as the first successful English colony in the New World—but a combination of bad planning and bad weather doomed the attempt. The winter of 1607–8, England's Great Frost winter, was just as severe on the other side of the Atlantic—"the hardest that ever was seen," according to one contemporary writer. The open bluff on which the Popham colonists planted their colony exposed them to the worst of the Atlantic gales. Popham himself died during the winter, and cold, hunger, and scurvy claimed the lives of many others. A contemporary account of the expedition notes that due to the "fear that all the other winters would prove like the first, the company would by no means stay in the country." And so, thirteen months after they had landed in Maine, the surviving Popham colonists sailed back to England, their ambitious and well-financed scheme defeated by the rigors of a Little Ice Age winter in North America.

The English had some justification (beyond hope) for expecting that the New World would be warmer than the old, for their geographers knew that New England lay considerably to the south of London—about 10 to 12 degrees of latitude south. Theories dating back to Aristotle and Parmenides of Elea, and still widely subscribed to during the age of exploration, held that the sole factor in determining a region's climate was the angle of the sun above the horizon. In fact, our word "climate" derives from the Greek word *klima,* inclination. The inclination of the sun *should,* by this theory, have given Boston about the same weather as Rome and Virginia about the same as the northern edge of Africa. What these European travelers to America did not reckon on was the climatic impact of the vast landmass to the west. The frigid winter northwesterlies that the early colonists complained of so bitterly blow across thousands of miles of frozen and often snow-covered ground, which serves to reinforce their chill. Largely ignorant of both American geography and weather patterns, the European newcomers advanced their own theories to explain the biting New England and Canadian winters. Marc Lescarbot, a French lawyer who spent some time at the early French colony at Port Royal (on present-day Nova Scotia) in the first decade of the seventeenth century, speculated that the winters were longer and colder there than in France because "the thickness of the woods and the greatness of the forests do hinder the sun from warming the ground . . . and besides that, this land having never been tilled is the more dampish."[12] In the following century European naturalists harped a good deal on these thick and "dampish" woods, attributing the supposed inferiority of animal and vegetable life in the New World to the "noxious vapors" that rose from them.

The New World colonies were not only colder in the winter than Western Europe, they were also hotter in the summer and, surprisingly, considerably wetter at most times of the year. American weather historian David Ludlum has assembled useful statistics comparing mean temperatures and rainfall levels in London, Boston, New York, and Norfolk, Virginia: Boston's January is more than 10 degrees colder than London's (29.2° F compared with 39.8° F), and its July nearly 10 degrees hotter (73.3° F compared with 63.7° F); London receives 24 inches of rainfall a year, whereas Boston gets 42.52 and Norfolk 44.58.[13] True, London was colder than today for long stretches during the Little Ice Age; but so was the United States.

For the first English settlers in Jamestown and Plymouth, the summer heat and winter cold and heavy rains and snows took some getting used to. "They that know the winters of that country know them to be sharp and violent, and subject to cruel and fierce storms," writes William Bradford of the Pilgrims' landing at Plymouth at the start of winter in 1620.[14] Two-thirds of the original Jamestown colonists died in the frigid winter of 1607–8. New words, for example "hurricane," adapted from the language of the Taino tribe of the West Indies, and "blizzard," had to be coined or imported into English to describe weather phenomena unknown in Western Europe. Early descriptions of hurricanes and violent thunder-storms convey some sense of how truly electrifying these weather events appeared to eyes accustomed to the mildness of Western Europe. "[It] grew not by degrees but came with violence in the beginning, to the great amazement of many," wrote William Bradford in his *Of Plymouth Plantation* of a hurricane that blew in from the southeast in mid-August 1635. "It caused the sea to swell to the southward of this place above 20 foot right up and down, and made many of the Indians to climb into trees for their safety."[15] More than a century later, in 1772, Alexander Hamilton, then a boy of seventeen living on the Caribbean island of St. Croix, sounds a similar note as he attempts to convey the horrifying power of a hurricane in a letter to his father: "Good God! what horror and destruction! It is impossible for me to describe it or for you to form any idea of it. It seemed as if a total dissolution of nature was taking place. The roaring of the sea and wind, fiery meteors flying about in the air, the prodigious glare of almost perpetual lightning, the crash of falling houses, and the earpiercing shrieks of the distressed were sufficient to strike astonishment into Angels . . ."

Even after the colonies weathered their precarious first years and eventually took hold and flourished, the colonists continued to look on the climates of the New World as something newly minted by God. This sense of freshness and innocence would persist and expand as the newcomers pushed westward over the next two and a half centuries into ever grander, stranger landscapes and ever more alien and extreme climates. The constant need to adapt to new climates may well have left its mark on the national character. So Robert Claiborne argues in his book *Climate, Man, and History:* "If empiricism, improvisation, the pragmatic determination to try anything once and keep trying as long as it works are hallmarks of the American character—and I am convinced they are—then the ever-new challenges of climate are surely an important part of the reason."[16] Indeed, one might say that this sense of the ever-new challenge, of the infinite promise, is the defining feature of America's weather. For European immigrants, this was not just a new world, but a cleaner, fitter, sounder, better world. "Here is an extraordinarie cleere and dry aire that is of a most healing nature," Francis Higginson wrote from the Massachusetts Bay colony in 1630. "A sup of New-England's aire is better than a whole draught of Old England's ales."

Despite its violent, sometimes deadly swings, America's weather eventually took its place in the imagination of the colonists as one more of her blessings. After only three years at the Plymouth Plantation, Edward Winslow wrote in his pamphlet *Good Newes from New England,* "I cannot conceive of any [climate] to agree better with the constitution of the English."[17] William Wood was even more sanguine about the climate of the Massachusetts Bay Colony in his *New Englands Prospect* of 1634: "it is for certain the best ground and sweetest climate in all those parts . . . agreeing well with the temper of our English bodies, being high land, and sharp air . . ." Wood admits that some may object that "it is too cold a country for our English men, who have been accustomed to a warmer climate," but he counters by pointing to the plentitude and cheapness of wood for building warm houses and good fires. America's cold, though "searching" and "piercing," is cleaner, drier, and healthier than the cold of England, for a New England winter "produces not so many noisome effects, as the raw winters of England." And, in any case, cold as a Massachusetts winter may be, it is, according to Wood, over in ten weeks at the most, "beginning in *December,* and breaking up the twentieth day of February . . ."[18]

Some of these early encomiums may have been propaganda or morale boosters: stuck as they were in the New World and exposed to its elements, the colonists might as well pretend to love it. But in a surprisingly short time, the English settlers of America ceased to think of themselves as English. Without coercion, they switched their allegiance to a land that was far vaster, richer, and more varied than anything they had known before and than any of them had dreamed. America's rough, strange, extravagant, unpredictable weather, as much as anything else, made these pioneers from Western Europe into Americans.

3.

COLONIAL AND
REVOLUTIONARY WEATHER

"When the Vapours rise and gather in thick Clouds, and the Lightning flashes with irresistible Power; let us then lift up our believing Eyes and see God in them . . ."

—Thomas Prince, "The Natural and Moral Government and
Agency of God in Causing Droughts and Rains," 1749.

"Surely the Thunder of Heaven is no more supernatural than the Rain, Hail or Sunshine of Heaven, against the Inconvenience of which we guard by Roofs & Shades without Scruple."

—Benjamin Franklin

ON THE AFTERNOON OF AUGUST 31, 1735, LIGHTNING STRUCK the meetinghouse at New London, Connecticut. It was a Sunday, a close and oppressive day of rest, and the faithful had spent the morning in church, as usual, under the direction of their pastor, Eliphalet Adams. The congregation broke for the midday meal and then returned to church for afternoon service, as was the custom in that deeply devout time. It was then that a storm whipped up off Long Island Sound. Towering black thunderheads massed in the sky. The Reverend Adams had to preach louder and louder to make himself heard over the rumbling crescendo. Gusts of wind bowed the trees and turned up the pale undersides of their leaves as the squall line approached. And then, in an electrifying instant that none of the New London parishioners would ever forget, a bolt of lightning connected with the church steeple. The effect could only be compared to the explosion of a bomb—but a bomb far more powerful than any device humans had in their arsenal in 1735. The force of the current knocked many people to the ground; several were seriously burned, and one young man died. The afternoon service was suspended and the surviving New London parishioners returned to their homes stunned and chastened.

We know as much as we do about this particular storm because the Reverend Adams later preached a sermon about it, which he published under the cumbersome but arresting title: "God Sometimes Answers His People by Terrible things in Righteousness: A Discourse Occasioned by that Awful Thunder-Clap Which Struck the Meeting-House in N. London, August 31st 1735." It is a document highly revealing of the prevailing attitude toward weather then current in New England. Orthodox Puritanism was on the wane in the American colonies by the 1730s, but religion maintained its grip over the intellectual and imaginative life of the region. Spirit, grace, providence, divinity—whatever name one used for the power and the glory of the Protestant God, it infused every aspect of experience and creation, from the most trivial to the most transcendent. Weather wonderfully straddled the two. Weather, in its unpredictability, fickleness, and occasional majestic violence, was a perfect symbol of the Divine Will. And thunder and lightning, then as now, were among the more dazzling and inexplicable of heavenly displays. And so, when a lightning bolt struck the meetinghouse in New London, Connecticut, on a hot August afternoon, the Reverend Adams quite naturally took it as a Sign. He knew from reading the Bible that thunder is the Voice of God, and, as he told his congregation, "we should stand in aw [sic] of him when he thus loudly utters his Voice." The Lord's judgment was unmistakable on that stormy August Sunday: "For there was need of something very Awakening to Rouze us out of that deep security, into which many of us were fallen. . . . [I]t is not unbecoming Even godly people to stand in aw of God's terrible Majesty at such a time . . ." Since the Puritans, like the Jews, considered themselves a chosen people who had entered into a special covenant with God, violent weather struck them as the very handwriting of God: whatever transpired in the heavens was a direct communication from on high, with a special significance for them and them alone. The message of the August 31 storm was clear and inescapable: the Lord wanted to blast the people of New London out of complacency. Only a sinner or a fool would try to evade this dread meaning.

Adams's sermon is not all fire-and-brimstone moralizing. A note of naked human shock enters in when he tries to re-create the horrible moment when the lightning hit the church roof: "the fire of God fell among us, rending the House in a wonderful manner . . . Striking down to the ground great Numbers of people, Burning and wounding several." Of the youth who died, Adams said that the lightning "licked up his spirits beyond recovery . . . All of us [were] Stunned and astonished with the sudden and surprizing Blow. Methinks the dreadful Sound is still in our Ears and we shall scarce forget it to our Dying Day." There was no thought of continuing the service after the storm had passed, for "the Lord himself was pleased to preach to us more immediately, that afternoon in a tremendous manner."[1] "We shall scarce forget it to our Dying Day": how many times have we heard these words, or others meaning the same thing, uttered in shock by those who have survived a brush with severe weather? No matter whether one explains a killer storm as God's tremendous preaching or as atmospheric turbulence created by the collision of two air masses of sharply different temperatures, the terror it arouses is impossible to shake.

Theological meteorology, as one scholar terms the Puritans' method of interpreting climatic phenomena, was the dominant approach to weather in New England from the landing of the Pilgrims at Plymouth during the stormy winter solstice of 1620 until the mid-eighteenth century. Drought, flood, severe cold, unbearable heat, deadly blizzard, and life-giving spring rain: these were all taken without question as the work of God—or the devil. Such notions were not merely figures of speech that preachers like Eliphalet Adams used to embellish their sermons and keep their congregations in line. The early settlers put their meteorological theory into practice whenever they could, praying and fasting for whatever kind of weather they desired, and presumably mending their sinful ways until their prayers were answered. William Bradford, governor and histo-

rian of Plymouth Plantation, gives a vivid description of what was undoubtedly New England's first Christian rainmaking ceremony, occasioned
by a prolonged drought in the summer of 1623:

> I may not here omit how, notwithstanding all their great pains and
> industry, and the great hopes of a large crop, the Lord seemed to
> blast, and take away the same, and to threaten further and more sore
> famine unto them. By a great drought which continued from the
> third week in May, till about the middle of July, without any rain
> and with great heat for the most part, insomuch as the corn began to
> wither away though it was set with fish, the moisture whereof helped
> it much. Yet at length it began to languish sore, and some of the
> drier grounds were parched like withered hay, part whereof was never
> recovered. Upon which they set apart a solemn day of humiliation, to
> seek the Lord by humble and fervent prayer, in this great distress.
> And He was pleased to give them a gracious and speedy answer, both
> to their own and the Indians' admiration that lived amongst them.
> For all the morning, and greatest part of the day, it was clear weather
> and very hot, and not a cloud or any sign of rain to be seen; yet
> toward evening it began to overcast, and shortly after to rain with
> such sweet and gentle showers as gave them cause of rejoicing and
> blessing God. It came without either wind or thunder or any vio-
> lence, and by degrees in that abundance as that the earth was thor-
> oughly wet and soaked therewith. Which did so apparently revive
> and quicken the decayed corn and other fruits, as was wonderful to
> see, and made the Indians astonished to behold. And afterwards the
> Lord sent them such seasonable showers, with interchange of fair
> warm weather as, through His blessing, caused a fruitful and liberal
> harvest, to their no small comfort and rejoicing. For which mercy, in
> time convenient, they also set apart a day of thanksgiving.[2]

This, of course, is the origin of the holiday we observe to this day, though
most of us are unaware that the first thanksgiving was a weather celebration.

Another weather ritual common in America throughout the colonial period was the ringing of church bells during electrical storms to ward off the "demons of the air." The custom had spread to America from Western Europe, where it dated back to the days of Charlemagne. As a guarantee of their meteorological potency, the church bells were inscribed with advertisements of their powers ("The sound of this bell vanquishes tempests, repels demons, and summons men," ran one such inscription) and consecrated at solemn ceremonies prior to installation. "Whan they shall rynge or sounde that bell," one sixteenth-century English bishop intoned as he washed a new church bell with water and salt and anointed it with holy oil, "all the disceiptes of the devyll may vanyshe away, hayle, thondrying, lightening, wyndes, and tempests, and all untemperate weathers may be aswaged."[3] How the bell performed this general cleansing of the atmosphere was left unexplained, but presumably it had something to do with its volume or the purity of its tone. The practice was astoundingly tenacious, persisting well into the nineteenth century despite the many well-recorded instances of God-fearing souls getting struck by lightning while in the act of ringing the consecrated bells.

FASTING AND PRAYER AND PREACHING AND BELL RINGING ARE pretty much what we would expect of Puritan "weathermen." If you believed, as they did, that weather rises from "the Breath of God," then why bother looking for more rational explanations for meteorological events? Superstition has an explanation for everything. But in fact, superstition alone did not satisfy the restless, quarrelsome intellectuals of colonial New England. The Puritan mind was surprisingly open to science. As Perry Miller writes in *The New England Mind*, "From the beginning, the Puritans had been hospitable to physical science. Obliged to study events in order to decipher the will of God, they welcomed the help of physics." Weather was, naturally, among the "events" that a predominantly agrarian ministry would need to study most minutely, and in interpreting these events, Puritan thinkers applied a curious blend of what Miller calls "natural causation" and "divine determination."[4] Crack the shell of their righteous rhetoric and you'll find a dense, wrinkled nut of scientific curiosity. Probably the most compelling case of this odd mingling of science and religion is offered by Increase and Cotton Mather, the father and son who dominated the Boston intellectual scene for half a century, from the 1670s through the 1720s.

Increase Mather, staunch upholder of the Puritan party line, pastor of Boston's North Church for sixty-two years, author of stacks of fiery sermons and scores of religious books, was fascinated by extreme weather. Dire meteorological events roll through his sermons. Thunderstorms filled him with an almost childlike curiosity. No matter how preoccupied he was with church matters or colonial politics, he would always make time for anyone who wanted to tell him a good, ripping story of calamitous weather. Blizzards, tornadoes, typhoons, comets, floods—he relished knowing the minute particulars of all climatic cataclysms. In 1684, Mather compiled the disaster stories he had heard over the years in a volume called *Remarkable Providences*—a collection of contemporary accounts of "remarkable" rescues at sea, outbreaks of witchcraft and demonic possession, sudden deaths visited upon the sinful, and breathtaking deliverances of the pious, along with suitable philosophical meditations. *Remarkable Providences* has fully three chapters devoted to remarkable weather. In the brittle old pages of this volume you hear the Puritan wind howl and the Puritan thunder crash—and you feel the terrific *tension* of Mather's effort to apprehend and explain these heavenly events without breaking the bonds of his faith. Reading *Remarkable Providences* is like witnessing a wrestling match between the natural and the supernatural, between reason and superstition. Mather's remarkable little book is the gusty, humid morning that heralds the advance of a vigorous cold front—in this case a front that ushered in the sprawling high-pressure system of the Enlightenment.

"There are who affirm, that although terrible lightnings with thunders have ever been frequent in this land, yet none were hurt thereby (neither man nor beast) for many years after the English did first settle in these American desarts," opens the chapter "Concerning Remarkables About Thunder and Lightning," "but that of later years fatal and fearful slaughters have in that way been made amongst us, is most certain. . . . Such solemn works of Providence ought not to be forgotten." And to ensure their long remembrance, Mather proceeds to relate the particulars of the most remarkable occurrences, as recounted to him by reliable witnesses —viz., in July 1654 one Mr. Partridge of Salisbury, "esteemed a very godly person," is struck dead by lightning, his clothes and skin perforated with holes "like such as were made with shot," though his shirt and body were scorched on one side only; on July 15, 1665, "terrible cracks of thunder" are heard in Boston and a householder returns to find the dishes melted on his shelves; one Captain Davenport, "a worthy man, and one that had in the Pequot war ventured his life," is killed by lightning while asleep, conclusively disproving Plutarch's "old opinion" that "men asleep are never smitten with lightning"; in May 1666 Samuel Ruggles, out on the road with a "loaden" cart, is struck by lightning and thrown ten feet away from his horse and oxen; though the animals are all killed, Ruggles himself escapes with only minor injuries—his right leg and part of one hand "stupefied" and all the hair on the leg "quite burnt off." And so on.

A take-heed-all-ye-sinners moralizing, similar to that of Eliphalet Adams's sermon, is implicit in all these remarkable occurrences: we tremble as we read that we might be next in line to be struck dead by a "fireball" while sleeping. But underneath the heavy bass of religion Mather sounds the clear, piercing note of rational inquiry: Why were these particular people singled out for death by lightning? What do their injuries and damages reveal about the nature of lightning? What is the relationship between thunder and lightning? Why does the same bolt of lightning kill some and merely stupefy others? What accounts for the recent outbreak of "fearful slaughters" in New England after so many years of terrible but harmless storms?

Mather sifts and compares the evidence at hand for his answers: the "brimstone" smell that lightning leaves behind makes it "manifest that these wonderful meteors are generated out of a nitrous and sulphurous matter"—and not a "bolt of stone" as others have claimed. And furthermore, "the nitre in the lightning may likewise be esteemed the natural cause of its being so penetrating and burning a nature." Since several victims of lightning strikes died without apparent bodily harm, Mather concludes that lightning slays by consuming the "vinculum," which he defines as the "tye of union between the soul and body." A very "subtile" form of death. As for the question of how these deadly storms get going, Mather writes, "It is not heresie to believe that Satan has sometimes a great operation in causing thunder-storms," but "holy angels" have far more meteorological power than any devil. The rationalist in him, briefly seizing control of the pen, asserts that "both natural causes and angels do many times concurre" in making thunder and lightning; but then faith muscles in on reason to insist piously that "the supream cause must not be disacknowledged: the Eternal Himself has a mighty hand of providence in such works. He thundreth with the voice of His excellency." And so, flinging aside demons, angels, and natural causes alike, the great preacher proclaims that thunder is "the voice of the Lord" and lightning the "arrows of God"—and then caps his discourse with a fitting moral: "most miserable is the state of all Christless sinners, who know not but that every thunder-storm which comes may send them to hell in a moment." With that, Mather slams the book shut on thunder and lightning, takes a deep breath, and proceeds to a "remarkable relation" concerning witches in Connecticut.[5]

For all its stern Calvinist trappings and its halting steps toward rational inquiry, the theological meteorology of Mather and Adams had more in common with Native American mythology than the white colonists would have cared to admit: one can practically see the Indian thunderbird perched on the shoulder of the Puritan Lord as He "thundreth with the voice of His excellency." The Indians themselves, according to Mather, acknowledged the ascendancy of a divine rainmaker more powerful than their spirits: having witnessed how swiftly the Puritans could summon up a rainstorm through fasting and prayer, the "heathen, now for more than twenty years upon occasion of want of rain, will speak to us to call upon the name of the Lord our God."[6] Mather, however, fails to mention how much useful information the first white settlers received from "the heathen" concerning the rigors of the land and its fickle climate —when and how and what to plant, where to build so as to avoid the scouring northwest winds of winter and the floods of spring, how to "read" the weather signs of the New World animals and plants. Even the most devout New Englanders knew better than to count on fasting and prayer alone for the success of the crops. Natural causes might not stand a candle to "the supream cause" in the great Puritan scheme of the cosmos—but anyone whose livelihood depended on the weather, and in the seventeenth century colonies that took in just about everyone but preachers and politicians, knew that one ignored natural causes at one's peril. As Mather so acutely pointed out, "both natural causes and angels" did "concurre" in determining the temper of the skies over seventeenth-century New England—now violently stormy, now serene; now shot through with ominous signs and portents, now conducive to fat harvests and smooth sailing; now mystical, now empirical. Partly cloudy with a chance of revelation about sums up the long-range outlook for Boston and vicinity at the end of the seventeenth century.

Increase Mather's son Cotton extended this forecast into the first decades of the eighteenth century. Notorious today for the ponderous load of prose he left behind and for his strident justification of the hanging of witches at Salem, Cotton Mather was in fact a rather enigmatic figure—at once more conservative than his father theologically and more advanced scientifically. His sermons, like his father's, are peppered with references to the deity's absolute control over the weather, as here in "Winter Meditations" of 1693: "If God says to the Snow, and the Rain, and the Like, Be thou upon the Earth, straight way they are so." Or here, in his journal from the furiously snowy March of 1717: "Mar. 18. This a time of much rebuke from Heaven upon us, in the season. Such storms and heaps of snow, visit us in the approach of spring, as were hardly ever known in the depth of winter. . . . I would procure a day of humiliation and supplication to be kept upon these occasions."[7] And yet Cotton Mather, the complete Puritan intellectual, keenly followed all the latest advances in science and philosophy of his day, fervently admired Newton's *Principia,* and regularly reported on the natural curiosities of America to the Royal Society in London, to which he was elected a Fellow in 1713. A champion of the mutual compatibility of reason and faith, Mather believed that weather expressed "the immediate Will of our most wise Creator" while also adhering perfectly to the laws of science. To quote Perry Miller once again: "The natural world is therefore reasonable yet mysterious—and so infinitely susceptible of spiritualization."[8]

In the spirit of scientific inquiry, Cotton Mather wrote a vivid report for the Royal Society about the mountainous snows that beset New England in the late winter of 1717, a document which Henry David Thoreau, with characteristic acerbity, insisted was "worth all the philosophy [Mather] might dream of." Mather makes no reference to "divine determination" as he surveys the bleak white devastation around him—cattle, turkeys, and pigs buried under snowdrifts; women about to give birth "putt unto Hardships" when fast-accumulating snow rendered their streets impassable; cottages disappearing under snowbanks without even a trace of their chimneys; "Religious Assemblies" suspended "throughout ye Country"; apple trees split to pieces with snow crusted up over their branches. Mather notes that "Indians near an hundred years old, affirm that their *Fathers* never told them of any thing that equalled" this March snow. And yet as American history rolled on, several other memorable March blizzards were to rival the overwhelming snows of 1717: the famous blizzard of 1888 whited out New York City on March 13, and on the same date in 1993 the so-called Storm of the Century swept up the East Coast (see Chapter 6).

Thomas Prince, another outspoken Boston cleric and a follower of the Mathers, also strove mightily to yoke together natural science and orthodox Puritanism in explaining the volatile weather of New England. During the summer of 1749 a "most extraordinary, long, extream, destroying Drought" afflicted the Northeast and Prince preached a sermon on the subject at the summer's end. "The Natural and Moral Government and Agency of God in Causing Droughts and Rains," as Prince titled his address, opens on the familiar chords of the jeremiad: God brought on this "afflictive Drought," Prince explains, because of "the Wickedness of the People that dwell therein . . . ." and he won't end the drought until the sinners are "excited to humble themselves before him, repent of their Offences, return in their Submission to Him, and implore his Mercy." But after about half an hour of this, Prince suddenly changes gear and launches into a highly technical discourse on meteorology, touching on air pressure, the influence of the sun on cloud formation, and the typical behavior of the winds over land and sea. "The Atmosphere round the Globe, is continually in every Part, either expanding above or contracting with in it's [sic] Balance of Gravitation," Prince informs his parishioners, "and . . . this is a main, continual source of our Varieties of Winds and Weather." There is a long patch of scientific rumination in this vein before Prince wrenches the topic back to God.

When one reads the sermon today, the contradiction right at the heart of Prince's argument seems glaringly obvious and one wonders that it didn't trip him up: on the one hand he's saying that lousy weather is a divinely inflicted punishment and on the other hand that it's an accident of chaotic global wind circulation patterns. How can it be both? No problem. Faith, as usual, smooths out all wrinkles. As Prince reminds his congregation with thundering conviction: "We should not dare to apprehend that the Sovereign God in governing this lower World intirely confines Himself to act according to his common Course of Nature; but most wisely and justly reserves the Liberty of acting otherwise on all Occasions when he sees most fitting." In other words, God makes the laws of weather and then breaks them when he feels like it. He deliberately keeps us in uncertainty about the behavior of the weather as one of his "powerful Restraints" on our "evil Appetites," and as an "Excitation . . . to all Kinds of Duties." Weather in Prince's formulation is thus a carrot that God holds out to keep us moving on the right track and a stick with which he beats us when we stray.

Puritan meteorology could go no further than this. The mental strain of reconciling a fundamentalist religion with an increasingly sophisticated science was too great. In 1684, Increase Mather wrote, "An orthodox and rational man may be of the opinion, that when the devil has before him the vapors and materials out of which the thunder and lightning are generated, his art is such as that he can bring them into form."[9] But by 1749, when Thomas Prince preached on the agency of God in causing drought and rain, the orthodox and the rational were no longer cohabiting on such easy, conjugal terms. In the violent tug-of-war between the natural and the supernatural, something had to give. What gave, of course, was the supernatural. It toppled over backward when a new player threw his weight onto the side of the natural—a writer, thinker, printer, tinkerer, inventor, and amateur scientist by the name of Benjamin Franklin.

WE DON'T KNOW EXACTLY WHICH DAY IT WAS, BUT WE KNOW it was June and we know it was 1752 when Ben Franklin and his son stood at the edge of a field on the outskirts of Philadelphia and flew a kite into the teeth of a thunderstorm. Franklin had been obsessed with lightning for years—so obsessed that four years earlier he had given up his lucrative printing business (which generated half of his income) to pursue his research in relative peace. Peace yielded illumination. After pondering the workings of the Leyden jar (a glass bottle equipped with wire and cork that could be charged with electricity—the original capacitor) Franklin became convinced that lightning is electricity—a flash of genius that extinguished centuries of superstition—and by the late spring of 1752 he had hatched a way to test his daring hypothesis. And so, on that now legendary day in June, when the wind kicked up and the black clouds crowded out the hazy Philadelphia sun, Franklin and his son grabbed their kite, their twine, their Leyden jar, and their hats and hurried off to the out-of-the-way field which they had chosen as the site of their secret experiment.

Nowadays we have grown accustomed to weather fanatics chasing tornadoes in pickup trucks or shooting videos of hurricanes, but back in the mid-eighteenth century the sight of two grown men launching a kite into a thunderstorm would have raised eyebrows pretty high—if there were any onlooking eyebrows in the vicinity, which is yet another detail of the occasion that we do not know. But Franklin was willing to risk more than embarrassment in conducting this experiment. He had seen for himself what a single bolt of lightning could do when it connected with the roof of a house. He knew he was playing with wildfire and possibly courting death. In the event, most scientists agree that Franklin was incredibly lucky to have survived the triumphant moment of his scientific career.

The silken kite with a wire attached to one end, the twine with a key tied to it, the "Leyden jar": the ingenious, childlike simplicity of Franklin's apparatus makes the experiment all the more thrilling. Franklin described how it worked a few months later in a letter he wrote to a fellow enthusiast of electricity:

> As soone as any of the Thunder Clouds come over the Kite, the pointed wire will draw the Electric Fire from them, & the Kite, with all the Twine, will be Electrified and the loose filaments of the Twine will stand out every way, and be attracted by an approaching finger.
>
> When the Rain has Wett the Kite & Twine, so that it can conduct the Electric Fire freely, you will find it stream out plentifully from the Key on the approach of your Knuckle.
>
> At this key the Phial may be Charged, and from Electric Fire thus obtained, Spirits may be kindled, and all the Other Electrical Experiments be performed . . . & thereby the Sameness of the Electric Matter with that of Lightning compleatly demonstrated.[10]

Franklin's conclusive demonstration of the "Sameness of the Electric Matter with that of Lightning" was an epochal event in the history of weather in America—right up there with the discovery of the jet stream and the launching of the first weather satellite. "The 'Prince of the Power of the Air' tumbled from his seat," one Victorian historian rhapsodized in recounting how Franklin's experiment dealt a "death-blow" to the bell ringers of theological meteorology; "the great doctrine which had so long afflicted the earth was prostrated forever."[11]

Prostrate—yes; dead—no. There is no question that thinking people everywhere welcomed Franklin's breakthrough. Though the kite experiment dealt with only one meteorological phenomenon, it ushered in a fundamental shift in "doctrine" and attitude toward weather—at least in the minds of those who understood its implications. But of course not

everyone understood or wanted to believe in it. Superstition about the
weather remained more palatable to many, as indeed it does today. As I.
Bernard Cohen writes in his book *Benjamin Franklin's Science,* "While we
often refer to the eighteenth-century as 'the age of the enlightenment,' we
must remember that this phrase applies to the thought of leaders in the
realm of the intellect, but obviously not to the vast majority of people
. . . superstition was still strong in the minds of the unenlightened."[12]
Franklin himself commented acidly on the continued resistance of religious
leaders to fitting church steeples with lightning rods (which he had invented as an immediate practical outcome of the experiment), even though
steeples were notoriously attractive targets for lightning: "Unlearned men,
such as commonly compose our church vestries," can hardly be expected to
shed their superstitious aversion to lightning rods "when we see how long
even philosophers, men of extensive science and great ingenuity, can hold
out against the evidence of new knowledge that does not square with their
preconceptions."[13] And yet those who did install lightning rods were
charged with "presumption" for trying to countermand God's decrees.
John Adams recalled hearing "Persons of the highest Rank among us, say
that they really thought, the Erection of Iron Points [lightning rods] was
an impious Attempt to robb the Almighty of his Thunder, to wrest the
Bolt of Vengeance out of his Hand."[14] In other words, if the Lord meant to
hurl a lightning bolt at your home, then you're sinning if you direct the
charge harmlessly into the ground. The Reverend Thomas Prince, who had
preached only a few years earlier on God's agency in causing droughts and
rain, went so far as to blame Franklin's lightning rods for a severe earthquake that hit the Boston area in 1755, alluding to a widely held belief
that lightning bolts striking the ground could bring on an earthquake.
(William Bradford, a little more than a century earlier, speculated that a
"great and fearful earthquake" that hit the Plymouth colony in June of
1638, may have been caused by a succession of cold and moist summers
"subject to early and untimely frosts."[15]) Cotton and Increase Mather were
both dead by the time Franklin made his experiment; one can only wonder
whether they would have welcomed or condemned the breakthrough.

"Surely the Thunder of Heaven is no more supernatural than the

Rain, Hail or Sunshine of Heaven, against the Inconvenience of which we guard by Roofs & Shades without scruple," wrote Franklin in defense of his lightning rods (which he refused to patent, though he stood to make a fortune had he done so).[16] The banishment of the supernatural was key to Franklin's science: it was the idea that rational inquiry *could* and *should* be directed at the Thunder and the Rain and Sunshine of Heaven that was so revolutionary in its day. As Ernest Cassara writes in his book *The Enlightenment in America,* "The new explanation of the nature of lightning, for example, had a profound effect on man's basic belief. If, as Franklin demonstrated, lightning struck because of atmospheric conditions, and not, as theologians claimed, because of the direct intervention of an angry God, the deity took up an existence at least once removed from the workings of His universe."[17] Cohen voices a similar idea: "Franklin's work on lightning and his invention of the lightning rod took on a special value in an Age of Reason. All enlightened men and women were convinced that the advance of scientific discovery should result in the diminution of superstition and fear of natural phenomena."[18] Terrifying, punishing weather continued to hit America, but it had a different meaning after Franklin.

The same intense curiosity about the workings of the natural world led Franklin, about a decade before the kite experiment, to his other great meteorological breakthrough: the discovery that storms along America's East Coast move in a vast circular pattern from southwest to northeast, despite the fact that the surface winds blow in off the ocean from the east or northeast (whence the name nor'easter). Franklin made this deduction based on just a few scraps of information that came his way by chance: On October 21, 1743, he had planned to observe a lunar eclipse in Philadelphia when clouds streamed in from the northeast to blot out the night sky. At exactly the same time Franklin's brother, 300 miles away in Boston, viewed the eclipse in clear skies, although the storm did eventually overtake the city later that night. When the brothers subsequently compared weather notes for that evening, they found that the nor'easter reached Boston about four hours after it had hit Philadelphia. Franklin extended his data base on the storm track by gathering reports from travelers and by sifting through articles about stormy weather in newspapers from Georgia to New England. Putting all this information together, he concluded that East Coast storms originate in Florida or Georgia and move northeast at about a hundred miles an hour. The implications of his reasoning were, once again, breathtaking: Franklin, in essence, had posited simultaneous observation as the basis for weather forecasting—an idea, as we'll see, that was not put into practice until a century later with the opening of a national network of telegraph lines.

THE HISTORIAN HENRY STEELE COMMAGER ONCE WROTE OF the Enlightenment that Europe imagined it but America realized it. What he meant was that the philosophical underpinnings of the Enlightenment —the belief that all the world's mysteries could be explained through reason; the emphasis on natural laws, order and classification; the triumph of science and empiricism over faith, tradition, and superstition—were laid down in Europe by Descartes, Newton, Leibnitz, Locke, Rousseau; but that the Americans put European theory into practice—in Franklin's inventions and writings; in the evolving social institutions of the thirteen colonies; in their declaration of independence from England; in the democratic republic that the newly created nation devised (and then revised) to govern itself. Most of all, the Enlightenment in America was realized in the life and work of Thomas Jefferson. Jefferson in a sense *lived* the Enlightenment: the pursuit of reason led directly to his passion for fact, his obsession with order, his extraordinary openness to all branches of knowledge and experience, his determination to put his ideals to work in the utopian microcosm he created at Monticello.

The fact that Jefferson, in addition to being a statesman, farmer, architect, traveler, writer, educator, naturalist, geologist, botanist, anthropologist, was also a weather fanatic was more than a temperamental quirk, like his taste for fine French wine. For Jefferson, weather was the visible countenance of nature, and nature, as another historian writes *"was* America for Jefferson."[19] Weather was the common denominator among Jefferson's many and varied pursuits, the essential link between his practical, political, and scientific sides. As another historian writes, with the possible exception of agriculture, "no other scientific field held Jefferson's interest like meteorology, in which he appears to have made himself the best-informed American of his day."[20] It is no exaggeration to say that Jefferson's thoughts never strayed far from the weather, for we know from his diaries, records, and garden books that he made twice-daily weather observations throughout much of his adult life, and encouraged other prominent citizens to do the same. Through his tremendous influence on the political and intellectual life of the country—and on its very size and shape through the Louisiana Purchase, which more than doubled America's territory—Jefferson brought a rational approach to weather into the mainstream of American life and thought. Jefferson made America's weather as integral to the nation's emerging identity as her new political system.

JULY 4, 1776. AFTER THREE DAYS OF VIGOROUS, SOMETIMES acrimonious debate, the delegates to the Continental Congress in Philadelphia agreed to endorse the Declaration of Independence. Jefferson, of course, was in the thick of the action, for he had undertaken to draft the original document and he was determined to preserve as much of his prose intact as he could (as it turned out, the delegates ended up slicing off about a quarter of what he had written and they subjected what remained to a light edit). And yet, in the midst of this electrifying drama, Jefferson took the time to record the temperature in his diary (it was rather cool for July 4 in Philadelphia, the temperature standing at 68 degrees at 6 A.M. and reaching only 76 at 1 P.M.) and to note that he had paid for a very fine thermometer that day, reimbursing an instruments dealer named Sparhawk the not inconsiderable sum of £3 15s. for it. Four days later Jefferson laid out an additional £4 10s. to this same Sparhawk for a barometer. These weather instruments were destined to receive much use over the next fifty years.

At one level, the coming together of the thermometer purchase and the signing of the Declaration is a charming coincidence, one of those bits of whimsy that weather nuts love to recount with a smile. And yet it's tempting to speculate that for Jefferson himself there was a deeper significance to the fact that he had equipped himself with first-class new weather observation instruments at the precise moment when his country, with his own signal assistance, proclaimed its freedom. Believing as he did in the vital importance of systematically studying and recording each day's weather, Jefferson would have seen the day of independence as an occasion on which to wipe the meteorological slate clean and begin afresh with the best new instruments money could buy. New political era, new weather records. Jefferson was an extraordinarily faithful weather watcher, keeping his *Weather Memorandum Book* up to date from July 1776, until six weeks before his death in the summer of 1826.

Jefferson's zeal in following the weather was more than a habit: it was practically a duty that he took upon himself as an educated man and a leader. George Washington was also a zealous weather diarist, and for the same reason. For both of these Founding Fathers, scrupulous daily weather observations laid down a foundation of knowledge on which to build an explanation of the mysteries of nature in America. Since Washington and Jefferson were both farmers before they were statesmen, they may have started the practice in the interest of their plantations; but they continued it in the interest of their country. As historian Charles Miller has written, meteorology, along with the other branches of what was then known as natural history, was a subject "fit for an Enlightenment scientist, because it promised to display the perfection of created nature; for a planter, farmer, and country gentleman because a knowledge of it contributed to his livelihood and was always available for his pleasure; and for an American because it symbolized his nation."[21] Jefferson in particular understood the national implications of daily weather observations. He had a scheme, which unfortunately never came to fruition, to equip one person in each county with weather instruments so that observations of wind and temperature could be made simultaneously twice a day, at dawn and again at 4 P.M., throughout the country. His dream, as he wrote a colleague in France, was to "ascertain the climate of the United States and its probable causes." This project was derailed by the war, but had it been implemented it would have been a forerunner of the National Weather Service, operating on the same fundamental principle of simultaneous observation that the

NWS bases its forecasts on today—the principle that Franklin identified as the key to storm tracking back in 1743.

Jefferson actually did manage to put his ideas into practice on a very small scale, getting the Reverend James Madison (the president of the College of William and Mary, not to be confused with the country's fourth President) to make twice-daily observations at Williamsburg during six weeks in 1778 while he made simultaneous observations at Monticello. "Had the telegraph been in existence," writes a Victorian weather historian of this early experiment, "there is no telling what these acute-minded colonists would have attempted."[22]

It's tempting to enshrine Jefferson as a kind of early American demi-god, but he did have his limitations. In most of the scientific endeavors that seized his interest, Jefferson was more a collector of facts and artifacts than a theorist: rather than defining his questions or working up hypotheses to prove or disprove, he usually plunged right into the compilation of vast quantities of data and in many cases he never progressed beyond this point. "His formal record keeping is scarcely to be believed," writes Miller. And yet there was a method to his compulsive tabulations. Jefferson worked on the assumption that all information about the natural world would eventually be pressed into service to illuminate the mysteries of creation: if only we *knew* enough, ultimately we would *understand.* This faith that nature *could be* explained through observation, classification, and analysis was one of the central tenets of the Enlightenment. It was Jefferson's greatness that he had the vision to project this faith on a national scale. The massive national projects that Jefferson sponsored or instituted —the Land Ordinance of 1785 (the federal act that imposed a perfect grid of six-mile squares on all the United States territories west of the Ohio River); the dispatching of Lewis and Clark to explore, map, chart, and describe the vast territories of the Louisiana Purchase; the never-realized dream of a national weather observation network—all were the expression of a single desire to launch the new republic according to the dictates of reason. In his passion to "ascertain the climate of the United States and its probable causes," Jefferson began with what was happening outside his window at Monticello and ended only at the Pacific Ocean.

JEFFERSON HAD A MORE URGENT MOTIVE THAN SCIENTIFIC CU-
riosity impelling him to "ascertain the climate of the United States," for at
the end of the eighteenth century our climate was under attack. European
men of science, especially in France, argued that New World animals
(including human beings) were stunted, deformed, few in number, and
short-lived because the Western Hemisphere had an inferior climate—
inferior to Europe's, of course. (It should be noted that the term "climate,"
as used in the Enlightenment, encompassed not only a region's characteristic weather conditions but also its total physical environment—topography, latitude and longitude, the influence of oceans, mountains, and rivers,
the quality of the air, the "improvements" people had made on the terrain
by clearing forests or rerouting rivers.) The European attack on America,
which grew out of the work of the great French naturalist Georges Louis
Leclerc, Comte de Buffon (1707–88), boils down to this: Cursed with a
climate both colder and wetter than Europe's, America (Buffon makes no
attempt to distinguish North and South America) is singularly "un-
friendly" to animal life. The New World is a hemispheric quagmire—a
region of dark, brooding forests choked with "noxious exhalations" and
vast swamps clogged with "stagnating waters." When subjected to the
unwholesomeness of this perpetually soggy, chilly, humid murk, "every
thing languishes, corrupts, and proves abortive. . . . [The earth in Amer-
ica] will never produce anything but humid creatures, plants, reptiles and
insects, and cold men and feeble animals . . ." Native animal species,
claimed Buffon, are runty and feeble compared with their Old World
counterparts, and even species introduced from Europe "shrink and dimin-
ish under a niggardly sky and an unprolific land." The human race is no
exception: "The savage," writes Buffon, "is feeble, and has small organs of
generation; he has neither body hair nor beard, and no ardor whatever for
his female . . . They have only few children, and they take little care of
them. Everywhere the original defect appears: they are indifferent because
they have little sexual capacity, and this indifference to the other sex is the
fundamental defect which weakens their nature, prevents its development,

and—destroying the very germs of life—uproots society at the same time."[23] If only "the savage" could catch a few more rays, he would take a more fervent interest in *l'amour*. Since Buffon never visited America, one can only marvel at his intimate acquaintance with the natives' organs of generation, sexual capacity, and mating habits.

Buffon's followers the Abbé Raynal and Corneille De Pauw went even further. De Pauw asserted in the late eighteenth century that the American continents are covered with "putrid and death-dealing waters" and the air is so "pestilential" that one could contract syphilis just by breathing it. Raynal and De Pauw argued that this pervasive miasma stunted the bodies and minds not only of the savages but also of the white settlers of the Americas. As De Pauw puts it, "Through the whole length of America, from Cape Horn to the Hudson Bay, there has never appeared a philosopher, an artist, a man of learning whose name has found a place in the history of science or whose talents have been of any use to others." Two continents of unrelieved mediocrity—and all because of noxious exhalations and stagnating waters! The Buffon theory of degeneracy, for all its lofty French rhetoric, sounds even more cracked than Increase Mather's supernatural explanation of lightning. But equally bizarre attempts to correlate climate and civilization would crop up again in later centuries.[24]

Jefferson shouldered the responsibility of defending his country and its climate from these foreign slurs. In a celebrated passage of his *Notes on the State of Virginia* (published in 1787), Jefferson summarizes the Buffon theory of degeneracy and then proceeds to demolish it point by point. He rounds up an impressive range of quadrupeds and compares the maximum weight each species has attained in Europe and America, conclusively demonstrating that everything from the beaver to the flying squirrel to the cow grows larger in the New World. Jefferson admits that European livestock sometimes grows larger and fatter than American specimens, but he asserts quite logically that this does not come about "by the *heat* and *dryness* of the climate, but by good food and shelter." Next he clears the name of the American Indian: he acknowledges that he has no knowledge of the Indians of South America and refuses to recount the "fables" published by others, but he knows from firsthand experience and reliable sources that the North American Indian is "neither more defective in ardor, nor more impotent with his female, than the white reduced to the same diet and exercise: that he is brave . . . that he will defend himself against an host of enemies . . . that he is affectionate to his children . . . that his vivacity and activity of mind is equal to ours in the same situation . . ." and so on.[25] Jefferson points out that even Buffon concedes that the American Indians are about as tall as the white Europeans. And so, he demands, "what inconceivable process" could have restricted the degenerating influence of the New World climate to the Indians' intellectual, moral, and sexual capacities? The only conceivable answer is: none. Finally, he turns to the supposed mediocrity of the white settlers of the New World. Here he lets the evidence speak for itself. America "though but a child of yesterday" has already produced "a Washington, whose memory will be adored while liberty shall have votaries"; "a Franklin, than whom no one of the present age has made more important discoveries"; and a Rittenhouse, "second to no astronomer living."[26] "We therefore suppose," he concludes, "that this reproach [of mediocrity] is as unjust as it is unkind; and that, of the geniuses which adorn the present age, America contributes its full share."[27]

Thomas Jefferson was not alone in rising to the defense of his country. Eulogies about the climate of the New World ran through the works of many American writers of the Enlightenment and persisted well into the nineteenth century, when Thoreau took it up. Ben Franklin slammed Buffon's notion that people in the New World were less fertile than their European counterparts: he pointed out that the white population of Pennsylvania was doubling about every twenty years, a phenomenon he attributed directly to "the salubrity of the air" in America and "the healthiness of the climate." William Barton in a paper delivered to the fledgling American Philosophical Society in 1791 made a virtue of the very climatic factors that Buffon had impugned: the frigid American winters "brace and invigorate the bodies of the people," making the North American climate the "most favourable to life." Timothy Dwight, a president of Yale College, noted that New Englanders loved their cold and snowy winters because travel by sleigh was "more convenient and more pleasurable" than travel by carriage over muddy, rutted roads, and thus there was "more visiting and more gaiety" in the winter. "The most tedious and disagreeable winters which we ever have are those which are warm and at the same time moist. Your countrymen who reside here inform us, that they are like the winters in England. To us it seems that nothing but habit can make them agreeable." Dwight goes on to conclude with rising regional fervor that New England, blessed with a "very pure" atmosphere and a "peculiarly bright" sky, is the healthiest country in the United States, and probably inferior in this respect to few in the world."[28] (He even marshals statistics—rather harrowing statistics—showing that an infant born in Europe has a "one to two" chance of surviving to age five, while a New England baby has a "one to two" chance of making it to seventeen.)

But Jefferson went further than anyone else, extending the climatic debate into the realm of the emotions, psychology, and mental health. America, he insisted, was not only as healthy as or healthier than Europe but far more cheerful. As he wrote his friend Constantin François de Chaseboeuf, Comte de Volney, the distinguished French philosopher, author, and scientist, in Europe "the sun does not shine more than half the number of days in the year in which it does in America"; America is "always clear" whereas Europe is "always cloudy"; and even though America gets twice as much rain as Europe, it falls in half the time. "Taking all these together," he concludes, "I prefer much the climate of the United States to that of Europe. I think it a more cheerful one." In another letter to Volney, Jefferson argues that the brightness and clarity of America's weather "has eradicated from our constitutions all dispositions to hang ourselves, which we might otherwise have inherited from our English ancestors." America's climate was thus ideally suited to the nurturing of a magnificent race of people—noble, free, robust, and *happy*. Europe could keep her mild gray depressing winters, her paltry drizzly rain showers, her tepid summers. In the New World, even the weather was new and clean and unencumbered: it was the weather, in short, of freedom.

The rebuttal of Buffon and the defeat of England in the Revolution were linked victories, engineered by the same enlightened leaders and arising from the same new nationalistic impulse. As historian Henry Steele

Commager writes in *The Empire of Reason,* the pressing need "to vindicate the [climate of] the New World from misunderstanding and calumny . . . encouraged Americans, for the first time, to consider their national, and their natural, character. It persuaded them to launch their own inquiries into Man and Nature in the New World. . . . In the past Americans had not thought of themselves as a people apart, but had taken for granted that they were Europeans, but now . . . they were becoming persuaded that they were indeed a separate people."[29]

Separate indeed, not only from their former masters in Europe but, as time went on, increasingly separate from each other. As more and more inquiries were launched into "Man and Nature in the New World," Americans became more sharply aware of their regional differences—differences that they attributed in no small part to climate and its effect on character. As the French farmer/intellectual St. John de Crèvecoeur wrote in his *Letters from an American Farmer,* "Europeans submit insensibly to these great powers [of climate, farming techniques, customs, and circumstance], and become in the course of a few generations not only Americans in general, but either Pennsylvanians, Virginians, or provincials under some other name . . . The inhabitants of Canada, Massachusetts, the middle provinces, the southern ones will be as different as their climates; their only points of unity will be those of religion and language."[30] No matter what state or region a citizen hailed from, he or she was convinced that it had the most delightful and salubrious climate in the Union. Jefferson harbored not a shade of doubt of the superiority of Virginia's climate to that of all the other states, and the superiority of Monticello to all other parts of Virginia. As he wrote his daughter Martha while he was shivering through a chill Vermont spring in 1791, "On the whole, I find nothing anywhere else, in point of climate, which Virginia need envy to any part of the world . . . When we consider how much climate contributes to the happiness of our conditions . . . we have reason to value highly the accident of birth in such a one as that of Virginia."[31] "There is not a healthier or more genial climate in the world," he wrote to Nathaniel Bowditch, trying to persuade him to relocate and teach at the University of Virginia.

Timothy Dwight, as we've seen, made the exact same claim for New England.

Jefferson's friend Volney, in his comprehensive volume *A View of the Soil and Climate of the United States of America,* based on his extensive travels through the new republic, writes perceptively of the bitterness with which residents of the North and South attack each other's climates and characters:

> [T]he southern people will terrify [an immigrant from Europe] from fixing in the north, by dwelling on their long and dreary winters, the hardships of excessive cold, the expence and apparatus which a bleak air and churlish soil occasion in living or cultivating the earth. The northern man, on the contrary, boasts much of his health, robustness, and activity, the gifts of labour, a sparing soil, and inclement skies; and rails against the pestilential bogs of the southern states, their sultry and incessant heats, their tormenting insects, the slothful and luxurious habits, and the crazy constitutions of the people, their gambling, drunkenness, and tyranny over their slaves: all produced by the very nature of their soil, and its luxuriant fertility.[32]

The mutual contempt between North and South, in which climate played a small but significant role, would explode less than half a century after Volney wrote these words.

Volney shrewdly points out that an individual's experience of climate and weather is largely subjective: one grows inured to whatever atmosphere one lives in for a long period of time, and takes it as the "norm" against which one measures all other climates. "Habit forms a sort of atmosphere around us," as he puts it, "to whose peculiarities our senses are dead, till we go forth and breathe a different air." Volney admits that he himself infinitely prefers the climate of Egypt, Syria, and the Mediterranean countries to all of the American states; however, if he *had* to settle in the United States, he would choose Newport, Rhode Island, the "southwest chain in Virginia, between the Roanoke and Rappahannock," or the shores of Lake Erie (though he would wait a hundred years for the "fevers" to be cleared off).

Volney's American translator, one C. B. Brown of Pennsylvania, could not sit idly by while his own state's climate was so unjustly defamed. In a series of amusingly pugnacious footnotes, he takes it upon himself to set the record straight on the *true* nature of Pennsylvania's climate. Brown dismisses out of hand Volney's assertion that the northeast wind in Philadelphia "oppresses the brain, and produces torpor and head-ache": admittedly the wet and cold blown in on this wind "are rather uncheering to the fancy," but they are certainly not "directly hurtful to the health." He heaps scorn on Volney's assertion that the summer heat in Philadelphia is "so intense" that "the streets are deserted from noon till five o'clock." Any native would smile at this, says Brown, for the truth is that we Americans are fully accustomed to "the great heats" of our climate and carry on our daily occupations no matter how hot it gets. "These principles operate beneficially," Brown concludes, "in spite of absurd modes in eating and dressing" (which, he points out, we have inherited from our ancestors in Europe) "and the almost general disuse of the bath. Vast numbers pass through a long life, amidst all these heats, clothed in cloth, flannel, and black fur hats, and lying on a feather bed at night, drinking nothing but wine and porter, and eating strong meats three times a day, and never allowing water to touch any part of them but their extremities, for a year together."[33] This is about as pungent a whiff of summer in the "good old days" as one could hope for.

The impulse to document the geography, natural features, climate, and history of the young United States was one that Volney shared with many other learned people of his day, both at home and abroad. Perhaps it was America's status as the first European colony to declare and win its independence; perhaps it was the opening up of the territories beyond the Alleghenies; perhaps it was the attacks of Buffon and his colleagues: any one or all of these sparked ambitious adventurers to wander to the farthest reaches of the American republic, amass as much factual information as they could, and then assemble it in voluminous compendia.

Jedidiah Morse produced one such volume, entitled *The American*

Geography, in which he takes up with a vengeance the notion that climate is the crucial determining influence on health. Morse painstakingly rates the climate of each state by its effects on character, health, and longevity: "Maine [is] very healthful. Many of the inhabitants live ninety years." Vermont: "None in the world more healthy. Snow begins to fall, commonly in the beginning of November, and is generally gone by the middle of April. During this season, the inhabitants generally enjoy a serene sky, and a keen cold air . . . Spring advances with surprizing quickness." In Rhode Island "the summers are delightful, especially on Rhode-Island [Newport], where the extreme heats, which prevail in other parts of America, are allayed by cool and refreshing breezes from the sea." Although consumption and dysentery are "prevalent" in this state, "these are not so much owing to the climate, as to intemperance and imprudence." Delaware, however, is unhealthy in many parts because "the land is generally low and flat, which occasions the waters to stagnate" (shades of Buffon). South Carolina is even worse, with "bilious diseases and fevers of all kinds . . . prevalent between July and October" in the low marshy sections of the state. But the heat, in Morse's view, is even more deleterious than the dampness of the low country, for it "relaxes the body" and engenders widespread "indolence," which is "probably a more operative cause in producing diseases."[34] Sir Augustus John Foster, an English gentleman who wrote up a breezy and gossipy account of his travels through Jeffersonian America, was even more alarmist about South Carolina's climate, insisting that "after the 10th June it is mortal to a planter to remain in the country near his rice grounds . . . [I]f they go even for a single day to visit the plantation, it is with imminent risk of life."[35]

WEIRD AS THEY SOUND TO US, THESE ATTEMPTS TO CORRELATE climate and character were all based on the soundest assumptions of Enlightenment natural science. Their authors believed that they were beating back ignorance and superstition through the exercise of right reason and scrupulous objectivity. They were convinced that the data alone, if amassed in sufficient quantities, would eventually cohere into the truth—hence the encyclopedic nature of so many of their scientific undertakings. Emotion, if present at all, must be buttressed with fact. Such was the climate of the late eighteenth century.

But as the Enlightenment waned, the American experience of weather was colored by a new cast of thought: an exalted Romantic reverence for the beauty and sublimity of meteorological phenomena. One hears the first notes of it in Crèvecoeur's *Letters from an American Farmer,* which was composed during the 1770s while he was farming in upstate New York: "A great thunderstorm; an extensive flood; a desolating hurricane; a sudden and intense frost; an overwhelming snowstorm; a sultry day,—each of these different scenes exhibits singular beauties even in spite of the damage they cause. Often whilst the heart laments the loss to the citizen, the enlightened mind, seeking for the natural causes, and astonished at the effects, awakes itself to surprise and wonder." Or here as Crèvecoeur stands with jaw agape, marveling at the spectacle of a blizzard in the Northeast: "How amazingly changed is the aspect of Nature! From the dusky hues of the autumnal shades, everything becomes refulgently white; from soft, miry roads, we pass all at once to slide on icy bridges. What could an inhabitant of Africa say or think in contemplating this northern phenomenon? Would not it raise in his mind a greater degree of astonishment than his thunderstorms and his vertical suns?"[36]

Listen to how a New Englander named William Tudor, Jr., wrings poetry out of geography as he contemplates the beauties of the North American winter in an address delivered to the Phi Beta Kappa Society in Cambridge, Massachusetts, in 1816:

❡ . . . [T]he effects . . . of our climate composed of a Siberian winter and an Italian summer furnish peculiar and new objects for description. . . . In our winters, we have the sun at the same altitude as in Italy, shining on an unlimited surface of snow, which can only be found in the higher latitudes of Europe, where the sun in the winter rises little above the horizon. The dazzling brilliance of a winter's day, and a moon-light night, when the utmost splendour of the sky is reflected from a surface of spotless white, attended with the most excessive cold, is peculiar to the northern part of the United States.[37]

Timothy Dwight strikes the same rhapsodic chords in his early-nineteenth-century description of a thunderstorm over New England:

> The thunder-storms of this country . . . are in most cases equally beneficial and delightful. An immense grandeur invests them during the time of their approach. They are so frequent in ordinary seasons as to furnish an ample supply of rain for the demands of vegetation. The wind which brings them, and which blows one or two days after they have passed over, is remarkably pure, refreshing, and healthy. The earth, particularly in the months of May and June, the richest season of vegetation, is beautiful beyond description. The verdure glows with new life—the flowers exult with additional beauty and fragrance . . . The sun, in the mean time, beams through the purified atmosphere with a peculiar splendour. One, and often two rainbows are finely painted on the reverse of the departing storm. The clouds in the western regions, wrought into the boldest figures, and tinged in different places with all the elegant hues of the prismatic image, present to the eye mountains with summits of gold, and precipices of fire.[38]

This is a far cry indeed from the angels and demons and flaming divine vengeance of the Mathers—and also from Ben Franklin and his electrical fluid. Dwight's mingling of national chauvinism and exalted nature worship heralds the onset of the Romantic, a movement which came into full flower a generation later in America in the writings of Emerson, Thoreau, and their colleagues in Concord, Massachusetts, as well as in the dazzling meteorological displays of the American Luminist painters. For the writers and artists of American Romanticism, weather was the driving force behind the majesty and mystery of nature.

• • •

THE CONVICTION THAT AMERICA'S CLIMATE IS FUNDAMENTALLY *different* from Europe's runs deep in our history—starting with the first white explorers and persisting through the *philosophes* of the Enlightenment and the Romantic nature worshippers of the nineteenth century. Another theme that crops up surprisingly early and keeps cropping up again and again in our history is climate change, the notion that America's weather patterns have shifted from what they used to be and that the white newcomers are somehow responsible for this shift. Here is Cotton Mather discoursing on a national warming trend just four generations after the Pilgrim Fathers landed at Plymouth Rock:

> Our own Winters are, observably as Comfortably Moderated since the Land has been Peopled, and Opened, of Late Years. Our Snows are not so Deep, and Long, since the Progress that has been made, in the Clearing of our Woods; and our Winds blow not such Rasours, as in the Days of our Fathers when the Hands of the Good Men would Freeze unto the Bread upon their Tables, and the strongest Wine there would in a few minutes be hardly to be swallowed for its Congelation: yea, Water Cast up into the Air, would be Turned into Ice e're it came unto the Ground.

Scholar that he was, Mather rounds off his observation with a quote from Petronius: "As our Land grows better Cultivated, we shall have less winter and less Anger too among us."[39]

Jefferson, nearly a century later, makes exactly the same point, using some of the same phrases, in his *Notes on the State of Virginia:*

> A change in our climate however is taking place very sensibly. Both heats and colds are become much more moderate within the memory even of the middle-aged. Snows are less frequent and less deep. They do not often lie, below the mountains, more than one, two, or three days, and very rarely a week. They are remembered to have been formerly frequent, deep, and of long continuance. The elderly inform me the earth used to be covered with snow about three

months in every year. The rivers, which then seldom failed to freeze over in the course of the winter, scarcely ever do so now. This change has produced an unfortunate fluctuation between heat and cold, in the spring of the year, which is very fatal to fruits.[40]

Volney concurs that the clearing of fields has led to a rapid and widespread warming of North America's climate and he asserts that the exact same thing took place in Europe, and "probably in Asia, and every where in the old world, history representing the climate of all countries as colder formerly than at present." And he quotes from an early history of Vermont: "When our ancestors came to New England, the weather and seasons were uniform and regular; the winter set in about the end of November, and continued till the middle of February. During this period, a cold, dry, and clear atmosphere prevailed, with scarcely any interruption. Winter ended with February, and spring came suddenly upon us. . . . Summer was extremely hot while it lasted, but it was generally limited to six weeks. Autumn and September began together, and the harvest was stored before the end of the month. The scene has greatly changed since that time, in the cultivated part of the country: the seasons are different, the weather more variable, the winter become shorter, and interrupted by great and sudden thaws. Spring is a scene of continual vicissitude, and these changes of temperature are extremely hurtful to vegetation. Summer is not so hot, but it lasts longer. Autumn is most tardy in beginning and ending, and the harvest is scarcely finished before the second week in November; nor does winter become settled and severe before the end of December."[41] *The seasons are different.* In the absence of reliable records we'll never know the facts of the matter. But in a way, the facts are irrelevant: it's the tenacity of the conviction that I find striking.

There is an undertone of anxiety in these early musings on climate change, and of guilt as well over the transformation that had been inflicted on the face of the Americas. The European whites christened these western continents the New World and celebrated their virgin vastness. But virginity is perishable. F. Scott Fitzgerald conjures up the electrifying instant of deflowering in the unforgettable final pages of *The Great Gatsby:* "for a transitory enchanted moment man must have held his breath in the presence of this continent, compelled into an aesthetic contemplation he neither understood nor desired, face to face for the last time in history with something commensurate to his capacity for wonder."

When the enchanted moment was over, the white colonists fell greedily upon the land and, like gods, began to make it over in their own image. For all their angry rebuttal of Buffon, Americans have been bent from day one on draining the "stagnating waters" of their wetlands and clearing the dark, dreary forests with their "noxious exhalations." Fear and loathing of forests run deep in American life. William Bradford describes how horrified the Pilgrims were by the first view of Massachusetts—"the whole country, full of woods and thickets, represented a wild and savage hue." An early traveler wrote of the American forests that "their deep and impervious gloom resembles the silence and solitude of death." Isaac Weld in his *Travels* noted that Americans "have an unconquerable aversion to trees" and that it was standard practice to clear-cut all the forest around a settlement. Ira Allen, patriot Ethan's brother, praised Vermonters for leveling large tracts of the state's forests: "the stagnant air vanishes with the woods, the rank vegetation feels the purifying exhalations flit off on lazy wing, and fevers and agues accompany them." In the fiction of Nathaniel Hawthorne the dark brooding "heathen wilderness" is the abode of the devil—a place to be shunned and burned.

And so we pushed back the woods as fast and as far as we could—and then we trembled at our hubris when we beheld the scalped landscape. Surely the climate change that we had brought on was but the outward sign of some immense disruption of the cosmic order. Surely we would be punished for tampering with so magnificent an entity, for corrupting so much innocence so quickly. American weather was wild and fluctuating and unpredictable when the Europeans arrived: but what if the weather that we created by carving up the land was worse? Volney believes that one early sign of the deleterious effect of climate change is the spread of "bilious fevers [which] have every where followed the cutting down of the woods, the clearing of the lands, and the drying of the swamps."[42] "Popular anxiety about climate change was, and still is, an ever-present component of the US weather culture," writes Andrew Ross in his recent book *Strange Weather,* "in a country whose geographical make-up (a large landmass that lacks any mountain ranges running horizontally across the country) produces a diversity of climates and a range of temperature extremes unmatched in any other nation-state."[43]

In these early theories of climate change there is also a strong element of mythologizing the past—the good old days of "our ancestors" when men were men and cold was cold and wine froze solid on the rough pine table. Mather and Jefferson worry that the warmer, gentler climate will end up making Americans soft and fat and mild—or corrupt and luxury-loving like the Europeans. Of course, the essential thing about the good old days of climate is that no one *really* remembers what they were like. As Timothy Dwight pointed out back in the early nineteenth century, "Few persons remember the state of weather for any length of time. Few, if any, registers were kept in former times; and few (all of them imperfect) have been published, except from week to week in newspapers, or other fleeting vehicles. Hence the comparisons of our present climate with that of former periods must be extremely defective."[44] Defective and subjective. No matter where you stand in time in America, the climate of preceding generations seemed to be far rougher, snowier, and more bracing than the present —right down to today. We have more science and more statistics at our fingertips to prove that climate has changed in the past century and that it may continue to change radically in the next century—but I'm convinced that the reason we worry about global warming is not very different from Mather's.

Dwight, a devoted weather watcher, was an outspoken dissenter from the warming theories of *his* day. "The winters of 1780, 1784, 1788, and 1805, were probably as severe as those of 1641 and 1696," he writes; "and the snow, which fell in 1717, was, I am persuaded, not so great as that which began to fall on the 20th of February, 1802. In 1641, and 1696, sleighs and sleds crossed the harbour of Boston, and some of them went down on the ice to Nantasket, nine miles. In 1780, the British dragoons passed from New-York to Staten Island, a distance of ten miles." Indeed, weather historian David Ludlum reports that the famous "hard winter" of

1779–80 was the only one in American history when all the waters surrounding New York City remained frozen over for weeks at a time. It's odd that when we think of weather in the American Revolution, we usually think of Valley Forge, Pennsylvania, where George Washington's men suffered through the winter of 1777–78. Some writers even point to that winter as a symbol of the Little Ice Age. But in fact the winter of 1779–80 was far more severe, not only in the northern states but as far south as Georgia and even down to New Orleans.[45]

Dwight reports that hard winters, though fairly common during his lifetime (1752–1817), were by no means the absolute rule. Mild weather, he notes, often persisted in Connecticut through the end of December and he has sometimes gathered "fresh-blown flowers" from his garden on Christmas Day. Hardly what one would expect at the tail end of the Little Ice Age. The summers, however, were markedly and consistently cooler than today: in the fourteen years of Dwight's records, the temperature rose above 90 only four times, all during the scorching summer of 1798. These days it's rare for *any* summer to pass without at least a week of 90 degree weather in southern New England.

A couple of memorably harsh winters in the first decade of the nineteenth century silenced much of the speculation about continental warming. But the cool summers that Dwight and others wrote about were far more remarkable, and even alarming in some years. Six or more inches of snow fell from Ohio and Kentucky eastward into Washington, D.C., Pennsylvania, New York, and southern New England during the first week of May 1802. Dwight reports seeing snow in July during the chilly summer of 1804. Twelve years later in 1816 came the notorious "year without a summer," during which temperatures fell below freezing every month of the year in New England and the mean temperature in Boston was $5\frac{1}{3}°$ F below normal. The numerous contemporary diaries and newspaper accounts have enabled weather historians to assemble a fairly detailed picture of this extraordinary season. After a backward spring, farmers rejoiced at the onset of hot summer weather in early June, with the temperature hitting 83 at noon in Boston on June 5. But then a surge of arctic air whipped through, plunging the temperature by 40 degrees in a matter of hours. On June 6, Boston temperatures never rose higher than 44, and snow fell on and off that day. The chill penetrated south to Virginia. A severe frost hit on June 10, and ten days later, at the summer solstice, the corn, beans, and cucumbers lay blackened in the fields. A contemporary observer noted that the leafless trees appeared "burned or scorched for miles" throughout the Northeast. The cold killed thousands of birds and in the north country newly shorn sheep died in vast numbers.

A milder interval (with temperatures going as high as 99 in southern New England on June 23) came to an end when another frost arrived on July 9, and there was frost again on August 22 and 29 as far south as Pennsylvania. Conditions were even more severe in northern New England. "The most gloomy and extraordinary weather ever seen," wrote Benjamin Harwood, a Bennington, Vermont, farmer, in his diary for June 6, 1816. "The awful scene continued," he noted glumly on June 8. "Sweeping blasts from the North all the forepart of the day, with light snow squalls."[46] Five inches of snow was measured in northern Vermont. "This is beyond anything of the kind I have ever known," wrote a New Hampshire weather observer. And in Maine, Joshua Whitman, a farmer in the western part of the state, wrote, "I presume the oldest person now living knows of no such weather [as] the 8th of June."[47] "This past summer and fall have been so cold and miserable," wrote weather diarist Adino Brackett, "that I have from despair kept no account of the weather." "The very singular seasons of 1816 will long be remembered," concluded the fledgling *North American Review,* one of the first American journals to publish meteorological data. The chill was not limited to the United States: Europe also suffered through a cold summer in 1816, and harvests were so meager that famine struck England, France, Germany, and Ireland (where hunger and the accompanying typhus epidemic killed 65,000 that year). Famine was averted in New England only because the wheat, rye, and oat harvests were abundant despite the summer snow, frost, and drought. But families dependent on corn were reduced to foraging in forests for roots, berries, nettles, and hedgehogs.

Scientists of the day believed that sunspot activity had caused the unseasonably chilly summer, and in fact sunspots were so pronounced during May and June of 1816 that people could observe them through smoked glass.[48] But most meteorologists today explain the record-breaking chill as a combination of a Little Ice Age climatic shift and the vast quantities of volcanic dust thrown into the atmosphere by a series of eruptions that occurred between 1812 and 1817—Soufrière on St. Vincent in 1812, Mayon in the Philippines in 1814, and the massive explosion of Tambora on the Indonesian island of Sumbawa in 1815, probably the most powerful volcanic blast of the past 10,000 years. Tambora's explosion killed all but twenty-six of the 12,000 Sumbawa islanders and threw an estimated 37 to 100 cubic miles of dust, ashes, and cinders into the atmosphere. The outpouring from Tambora spread throughout the earth's atmosphere, veiling the sun's rays in a reddish haze and blocking enough of its radiant energy to lower temperatures. Ben Franklin may have been the first to connect volcanic eruptions with chilly weather: back in 1785 he wrote of a "dry fog" filtering out the sun's rays after the powerful eruptions of Mount Asama in Japan and Skaptar Jökull in Iceland. William Humphreys, a scientist working at the U.S. Weather Bureau in the early twentieth century, carried out a formal study of Franklin's theory. According to Humphreys, the volcanic dust acts as a shield, screening out enough of the sun's rays to lower temperatures a degree or two—a sufficient change to bring on a year without a summer. We all had a taste of "volcano weather" recently, when Mount Pinatubo in the Philippines blew in 1991, triggering a series of chilly summers over most of the country (though not as cold as 1816).

One immediate effect of the cold summer of 1816 was mass abandonment of the luckless farms of New England and migration to the Midwest. The Zanesville, Ohio, *Messenger* reported on October 31, 1816, that "the number of emigrants from the eastward the present season, far exceeds what has ever before been heretofore witnessed." The reporter admits that "it is impossible to calculate to a certainty the number of persons," but gives a round estimate of "some thousands." We do know that Indiana gained 42,000 settlers in 1816. Weather historian Patrick Hughes notes that the "first general migration from New England to the Middle West" occurred in 1817—immediately following the year without a summer—

with the northern states of Maine, Vermont, and New Hampshire losing the most people. Nearly twice as many people migrated out of New England in 1816–17 as in the remainder of the decade. A team of Maine historians discovered after sifting through local records that this pattern of "out-migration" continued throughout the nineteenth century whenever the summers in Maine were unusually cold and wet. Dreams of prosperity died bitterly amid the blackened crops after yet another July frost or miserably wet August. Newspapers fanned the flames of discontent by griping about the "uncommonly cold and wet" conditions, fostering the idea that the weather was somehow to blame for not behaving "properly."[49] As we'll see, the exact same thing was to happen on the Great Plains a few generations later, though in this case it was heat and drought instead of cold and wet that laid waste to hopes raised on false expectations of climate.

Thousands migrated in search of more felicitous weather. Thousands more stayed home and prayed. "The alarm and depression so wrought upon the feelings of the community, that a religious revival ensued," noted a local historian of Madison County, New York, of the "cold season" of 1816. In Burlington, Vermont, an impressive new church at the head of Church Street was dedicated the following year, and a local tradition has it that parishioners prayed at the dedication that God would inflict no more years without summer. The conviction that weather was controlled by supernatural powers, either Christian or pagan or both, was deeply rooted in our national consciousness. Though from the nineteenth century on, the history of America's weather would be increasingly intertwined with the history of science, prayer and weather superstitions persisted. We have never really wanted to surrender our faith that we *can* do something about the weather, especially when the weather turns ugly. The voice of Increase Mather thunders on.

4.
WEATHER IN THE WEST

THE HISTORY OF WEATHER IN THE WEST IS A HISTORY OF continual revelation and revision. We Americans, practically from the first, have staked our hopes and dreams on the boundless western expanse of the continent. Yet we have never fully come to grips with the radical, elemental difference between the West and the original Atlantic seaboard colonies. In weather as in terrain, the West might as well be another planet. The dryness. The extremities of temperatures. The tornadoes and swift fierce thunderstorms. The "blue northers" that howl down from Canada to Texas, plunging temperatures by as much as 50 degrees in a few hours. The ceaseless winds of the Plains. The blinding blizzards. The summer flash floods that can turn dry creek beds and canyons into Niagaras within minutes. The enormous swings in precipitation from year to year. The parallel ranges of tremendous, jagged mountains that drain all moisture from the Pacific-born clouds and cast immense rain shadows in their lee. Nothing in the East, nothing back in Europe, prepared Americans for this environment. We have literally had to teach ourselves to see the West with new eyes, and as we have learned we have made some terrible, tragic mistakes. Our hopes and dreams and eastern expectations have gotten in our way again and again. Great American Desert or the Nation's Breadbasket? Eden or Dust Bowl? The very terms we have devised for the West reveal our eastern orientation. One can only wonder how our vision of this region would have differed had the flow of discovery and settlement been from west to east, instead of the other way around.

We speak of "the West" as if it were a single vast entity, whereas in fact this region across the wide Missouri comprises several nations' worth of climates and topographies. The Rockies, Sierras, and Cascade Mountains scissor out their own patchworks of microclimates—deserts ascending into rain-freshened forests, belts of eternal snow gleaming on the triangular peaks, pocket tundras starved of sun and rain alike. On any given day in January, Bismarck, North Dakota, can be 30 below zero and dazzled by a sky of icy clarity while in Seattle, Washington, two days' drive due west, it's 50 degrees and drizzling gray. Coastal California tends to get most of its rain in the winter months in prolonged drenching downpours while in Arizona and New Mexico the rains come in brief summer thunderstorms. The distinctions and varieties of western weather go on and on. And yet there are unifying factors. Dryness is one: slice off the region's northwest corner and the west-facing mountain flanks and you're looking at arid country, deprived in most years of enough rain to keep a lawn green, let alone raise crops, without irrigation. Low humidity goes with low rainfall: the air in the West tastes lighter, the sky looks bigger, deeper, more remote from the earth. Violence and surprise govern the weather out West. Twenty straight days of clear skies and gentle breezes lull you into security and then *wham* a thunderhead boils up over the ridge and a wall of brown water roars down the canyon. Blizzards in October. The hottest, driest summer on record one year and devastating floods a couple of years later.

When I talk about "the West" in the pages to follow, I am referring to this abstract unity, this underlying commonality that binds these disparate climates together. Choose a day on your calendar and chances are there will be many more weathers going on in the West than back East. Yet each of these weathers will probably have more in common with each other than any of them do with what is happening in New York, Tennessee, Pennsylvania, or the Carolinas. When you look at the West through the eyes of its early explorers and pioneering settlers, as I do below, the shocking *differentness* of these strange new weathers hits you full in the face. That's what this chapter is about.

AT THE HEAD OF THE AMERICAN HISTORY OF WEATHER IN THE West stands Thomas Jefferson—which is odd because Jefferson himself never traveled farther west than the Shenandoah Valley, and that was only once and just to graze its eastern fringe. Yet the idea of the vast, vague western reaches of the continent loomed large in Jefferson's imagination. "[N]obody had read or thought more about [the West] than Jefferson," writes historian Donald Jackson in *Among the Sleeping Giants*. "For twenty years, long before he became president, Jefferson had been trying to send an expedition westward." From Monticello, the westernmost dwelling of any early American statesman, Jefferson had a view of the Virginia hills undulating softly and seemingly infinitely to the western horizon. And he knew that somewhere, some unfathomable number of miles distant, the continent ended at the Pacific Ocean. But what lay in between was largely a mystery—a mystery Jefferson was determined to solve. The literature on the West available at the start of the nineteenth century—accounts written over the past 250 years by trappers, explorers, missionaries, and merchants —offered unpromising glimpses of barren deserts, immense snowcapped mountains, and smoking volcanoes. The Spanish explorer Don Francisco Vásquez de Coronado recorded the first European impressions of America's West on an arduous journey from Mexico through New Mexico, Arizona, Oklahoma, and possibly up to southern Kansas in 1540. He returned empty-handed with the conclusion that the desert was uninhabitable—too hot by day, too cold by night, too burdened by snow in winter, and above all lacking in gold. Pedro Castañeda, Coronado's chief historian, wrote in befuddlement, "This country is like a bowl, so that when a man sits down, the horizon surrounds him all around at the distance of a musket shot." The Pacific coast also struck the first European explorers as unfavorable in geography and unpleasant in weather: Francis Drake (not yet knighted), sailing the *Golden Hind* along the Washington-Canada border in 1579, complained of "the most vile, thicke and stinking fogges" and "congealed rain" off the coast of Oregon and then, when he reached California in June, he wondered at how brown and barren the parched shoreline was.

Jefferson, however, was undeterred. The West was a tremendous void waiting for the bright clear light of Enlightenment science: it begged to be explored, mapped, described, and catalogued, and then opened to commerce, to the lucrative fur trade (which was then largely dominated by the French and English), and possibly to future settlement by small farmers. Already Jefferson's agrarian grid was spreading rapidly through the "Old Northwest Territories" of Ohio and eastern Michigan. It was only a matter of time before the American settlements reached the Mississippi River. Beyond, opening out north and west of New Orleans, lay the unknown region known as Louisiana. Acquisition of Louisiana would magnificently extend what Jefferson termed "the empire of liberty." Spain and France had been playing a listless game of tug-of-war with the territory for years. Why shouldn't Americans explore it? Indeed, why shouldn't America have it? The West. The West. It was a dream and a goad for Jefferson, as for the hungry generations that came after him.

Then Louisiana fell into Jefferson's lap and the dream collided with reality. Through a tangled skein of international politics, intrigue, finance, circumstance, and military strategy, Napoleon decided to sell the 800,000-square-mile Louisiana Territory—which stretched north and west from New Orleans to encompass most of present-day Louisiana, Arkansas, Missouri, Iowa, much of Minnesota, the Dakotas, Nebraska, Oklahoma, most of Kansas, and the sections of Montana, Wyoming, and Colorado that lay east of the Rockies, though the precise borders were unknown—to Jefferson for the round sum of $15 million. As one historian has written, "Napoleon had no idea what he had sold for $15 million, and Jefferson had no idea what he bought." But, as it turned out, the Louisiana Purchase was the bargain of the millennium. With "Louisiana," Jefferson had at a stroke of the pen more than doubled the size of the American republic. The vast, vague, mysterious West was now American soil—"a habitable idea," as essayist and novelist Jonathan Raban puts it. "What it lacked was contour lines, weather, soil, flora, mineral deposits, birds, animals, and Indian nations."[1]

Jefferson lost no time in arranging to have these particulars pinned down. Even before the Louisiana Purchase had been concluded, Jefferson had broached the idea of making a scientific expedition into the Far West with his twenty-eight-year-old private secretary Meriwether Lewis, a captain in the regular army who had grown up near Monticello in rural Albemarle County. Lewis readily agreed and he enlisted Lieutenant William Clark, who had been his commanding officer when he served as an ensign in the regular army some years back and had remained a friend ever since, to lead the expedition with him. Lewis and Clark quickly assembled an exploring party, got themselves outfitted, and made plans to carry out their orders—to travel up the Missouri River to its headwaters and from there to cross to the "Highlands" and proceed to the Pacific Ocean using "the best water-communication which offered itself." The momentous journey began on May 14, 1804, when the exploring party pushed off from north of St. Louis. The weather was damp, but improving, as Clark recorded in his journal: "Rained the fore part of the day. . . . I set out at 4 o'clock P.M., in the presence of many of the neighboring inhabitants, and proceeded on under a gentle breeze up the Missouri."

Among the many detailed instructions with which he sent Lewis and Clark into the great unknown Jefferson included the following: to observe and document "climate, as characterized by the thermometer, by the proportions of rainy, cloudy, and clear days, by lightning, hail, snow, ice, by the access and recess of frost, by the winds prevailing at different seasons, the dates at which particular plants put forth or lose their flower or leaf, times of appearance of particular birds, reptiles or insects . . ." These are precisely the climatic details that Jefferson himself kept track of religiously each day in his journals and garden books. Though Jefferson did not place the weather instructions at the top of the list, he did make them longer and more detailed than any other set of instructions given to the Corps of Discovery. When Lewis and Clark returned, the weather of the West was to be a mystery no longer.

Weather does, in fact, figure prominently in the journals that Lewis and Clark kept during the two and a half years of their resoundingly successful expedition. Soaking rainstorms and hard winds seem to have been the prevailing conditions through much of the journey. Clark observes on August 5, 1804, somewhere in northeastern Nebraska, that "thunder and lightning is not as common in this country as it is in the Atlantic states," a conclusion that would no doubt surprise most people who have lived through a full summer in both regions. The explorers offer a degree-by-degree account of the bitter cold at Fort Mandan, North Dakota, where they wintered over from November 1804 until March, 1805: 29 below on December 11 ("the ice in the atmosphere being so thick as to render the weather hazy and give the appearance of two suns," writes Lewis), 22 below on December 16, 45 below on the seventeenth, "excessively cold" on the night of January 10. By April, Lewis and Clark can write with the authority of firsthand experience that the wind blows "with astonishing violence" on the Plains due to the absence of trees. Lewis also remarks on how "remarkably dry and pure" the air is "in this open country . . . The atmosphere is more transparent than I ever observed in any country through which I have passed." In Montana that June the weather was inclement—with a cold northeast wind blowing on the sixth followed by steady rain the next day (though Lewis notes curiously that the moisture fails to penetrate more than two inches of the dense clay soil). "Wind *as* usual now," writes Clark on June 17, 1805, already accustomed to the incessant blowing of the Great Plains.

Once they got into western Montana and the Bitterroot Mountains of Idaho, Lewis and Clark were in territory that no white person had ever seen before. The terrain—and very often the weather—was, and still is, spectacular in this part of the world: after a flat eternity of grassland the front range of the Rocky Mountains thrusts up all at once into vertical forest, tricky winds are sliced and channeled by mountain ranges, storms boil up while your back is turned. Even today, crossing this countryside at 70 miles per hour in a heated/air-conditioned car, one is awestruck by the topography and climate. The really strange part about the Lewis and Clark journals is how *un-awestruck* they were or allowed themselves to be in print. The 40-below-zero winter of North Dakota, the mid-September snow at

Lolo Pass, the incessant winter rain on the Oregon coast: each new extreme of weather was duly recorded without much comment or exclamation by the two men from Virginia, who had surely never seen anything like it before. "I beheld the Rocky Mountains for the first time, with certainty," writes Clark on May 26, 1805, in central Montana, a few miles upstream from where the Musselshell River flows into the Missouri. "Those points of the Rocky Mountains were covered with snow, and the sun shone on it in such a manner as to give me a most plain and satisfactory view. Whilst I viewed those mountains, I felt a secret pleasure in finding myself so near the head of the—heretofore conceived—boundless Missouri." Inching up the steep treacherous flanks of those mountains in early September, Clark notes, "At dusk it began to snow; at 3 o'clock some rain. The mountains to the east covered with snow. We met with a great misfortune in having our last thermometer broken by accident. This day we passed over immense hills, and some of the worst roads that ever horses passed. Snow about 2 inches deep when it began to rain, which terminated in a sleet storm." "We are continually covered with Snow," reads an entry from a few days later, "indeed I was at one time fearfull my feet would freeze in the thin mockersons which I wore . . ."

That is about as carried away as the captains allow themselves to get. Reading the journals today, one comes away with an abiding impression of monotony, imposed order, and aridity. As Bernard De Voto writes in his classic *Across the Wide Missouri,* Lewis and Clark had "crossed a dry country, and that made an impression, as the fact that they had also crossed the Stony Mountains did not."[2] Or as Marc Reisner, historian of western water, puts it, "The country Lewis and Clark saw amazed, appalled, and enchanted them. Above all, it bewildered them. They had seen the western plains at their wettest—in the springtime of an apparently wet year—but still there were few rivers, and full ones were fewer. The sky was so immense it swallowed the landscape, but the land swallowed up the provenance of the sky. . . . Reading their journals, one gets the impression that Lewis and Clark simply didn't know what to think. They had never seen a landscape like this, never guessed one could even exist. . . . Louisiana, though penetrated, remained an enigma."[3]

The enigma was slowly unraveled by those who came after Lewis and Clark, though new layers of mystery and mythology accumulated with each successive wave of exploration and settlement. Gradually, over the course of the nineteenth century, the enigma of weather in the West became focused on a single element: rainfall (or the absence of it). Among the earliest chroniclers of western aridity was Zebulon Pike, himself an enigmatic figure with shadowy connections to the Aaron Burr conspiracy. Setting out from St. Louis even before Lewis and Clark had returned, Pike, at the age of twenty-seven, was assigned a party of twenty-three men and fifty-one Indians and instructed to travel from the Missouri River to the Arkansas River and to follow this river to its source. Initially, Pike's party traveled by boat, as Lewis and Clark had done; but after a month they traded boats for horses and made far swifter progress. In Kansas, the main part of Pike's party struck out westward toward the mountains, drawing near the 14,000-foot southern Colorado peak that bears his name. In March 1807 the Spanish intercepted Pike's party (one wonders how they discovered a single handful of men in this immensity). Pike insisted to his captors that he had lost his way while hunting for the Red River, but the Spanish were convinced that he was a deceitful American spy, and transported him under guard first to Santa Fe and thence to Chihuahua, Mexico. Pike would surely have been executed had the Spanish authorities not been concerned about drawing retaliation from Washington, and he was returned to United States territory later in 1807.

Whatever the true nature of Pike's expedition, it took him through thousands of miles of territory that no Yankee had seen or written about before—across the entire width of Kansas, up into the front range of the Rockies in Colorado, down through the Sangre de Cristo Mountains of New Mexico, and eventually across the lower third of Texas. Whether or not he was a spy, Pike was unquestionably a keen observer of topography and climate and a far more gifted prose stylist than either Lewis or Clark. In his journal, published in 1810, Pike broke the news to his countrymen that vast stretches of the territory west of the Mississippi and south of the Missouri River were desert—"tracts of many leagues where the wind had thrown up the sand in all the fanciful forms of the ocean's rolling wave, and on which not a speck of vegetable matter existed." Hot and parched in the summer; cold and bleak in the winter; treacherous and unpredictable in the mountains: such was the weather in the West that Pike recorded. It was Pike, also, who first called the arid lands of the High Plains the Great American Desert—a label that in itself had a profound impact on the perception of western weather. Americans in the humid East had no experience of deserts except in Bible stories and historical fictions. But the word set off an instant reflex in the American imagination: a desert, it was assumed, was a wasteland cursed with an unendurable climate, a region to flee through, as the Hebrews did in Sinai, or to perish in. Desert meant sand, heat, drought. It also meant wind: an almost constant flow of air rippling the grasses of the treeless plains, blowing with more force and more constancy than in any other region of the country save the coast. The wind averages ten to twelve miles an hour on the central and northern Plains and fifteen in the Texas panhandle, but during storms it accelerates to a deafening roar, bringing in blizzards on icy blasts or transmogrifying overnight to snow-eating "chinooks" that can raise temperatures more than 50 degrees and devour over a foot of snow in a few hours or shifting yet again to "northers" that plunge temperatures just as deeply and quickly. To an agricultural people, of course, desert most of all meant no crops. Pike's "discovery" of a Great American Desert, confirmed by Major Stephen H. Long during his expedition of 1819–20 and thereafter incorporated into maps and geography books, made the Louisiana Purchase into something of a mockery. A billion acres of Sahara.

Jedidiah Smith, another early rider of the range, added more detail to the great western canvas in his journal of his 1827–28 expedition. Remembered today as the first white American to travel overland to California and the first to find a central route between the Rockies and the Pacific Ocean, Smith wrote even more memorably than Pike of the merciless weather and "blazing sands" of the Great Basin. "The sun of this day arose on the parched waste," he recorded on June 25, 1827, somewhere east of the Sierras, "and it seemed to us that we were the most unhappy beings on which it poured its floods of light." Smith concluded after surviving the rigors of an arid summer on the Plains that "hunger can be endured more than twice as long as thirst . . . . [Hunger is] light in comparison to the agony of burning thirst." Smith's attitude toward the weather does, however, brighten up a bit once he gets to California. "The air was extremely pleasant from the effect of a gentle North Western Breeze that rose and sank with the rising and setting of the sun," he notes on the California coast near the San Joaquin River—"pleasant" being a word notably absent from his journal until now.[4] Smith here confirms rumors of the pleasures of California's climate that had been circulating eastward since the late eighteenth century. Back in 1789, Jedidiah Morse asserted in *The American Geography* that the climate of California "must be very agreeable" judging from "its situation," even though the deceitful Spanish Jesuits had claimed "for political reasons" that it was "unwholesome" and "barren." Morse clearly never set foot on the West Coast, but he passed on the odd fact that "in California, there falls in the morning a great quantity of dew, which, settling on the rose leaves, candies, and becomes hard like manna, having all the sweetness of refined sugar, without its whiteness." This was truly the climate of Eden. Jedidiah Smith, in his eyewitness account two generations later, finds a serpent lurking in the California paradise that Morse had failed to mention: the winter rainy season, which muddies the ground and swells the rivers for three months running (except when the rains fail during drought years). But the early arrival of spring comes as an unexpected blessing. Smith concludes that all in all, aside from the "superabundance" of winter rain, the climate and situation of coastal California combines "many advantages [and] is at the same time most delightful and pleasant"—a notion that has remained firmly lodged in the popular imagination ever since. So here was another facet of the western myth: the image of California as an island of climatic serenity utterly unto itself. For Smith and those who followed the "great highway" through the mountains that he opened, California gleamed as the golden reward for surviving the hellish weather of the Plains, the desert, and the Sierra.

Henry Nash Smith notes in *Virgin Land,* his brilliant study of the mythology and iconography of the West, that as the concept of a Great American Desert took root, civilized folks in the East began to view the region in alarm as a breeding ground for "nomads and brigands." The hostile climate would produce a hostile race—a tribe of mongrels who lived by hunting and trapping and who would in time fan out from the desert to prey on the peaceful agricultural communities to the east. For the sixty years between the publication of Pike's journal and the end of the Civil War, Americans pretty much wrote off the Great Plains as a nonplace with nonweather—or with weather and topography so fearsomely other as to be beyond recognition. For the quarter of a million Americans who emigrated along the Oregon Trail between 1840 and 1860, "the interior West was not a place but a way," as Wallace Stegner writes, "a trail to the Promised Land, an adventurous, dangerous rite of passage."[5] Adventure—in passage. Danger often in weather. As one historian concluded after studying some of the journals kept by emigrants along the Oregon Trail in the 1840s, "The brilliant lightning storms, high winds, and hail of gross proportions amazed and intimidated those unaccustomed to such violence from the heavens."[6] One of the more articulate of the amazed travelers of this route was journalist Edwin Bryant, whose 1848 best-seller *What I Saw in California* describes the trek he took West two years earlier. Here is Bryant's account of a memorably ferocious night in north-central Kansas: "May 27.—A terrific thunder storm roared and raged, and poured out its floods of water throughout a great portion of the night. . . . the whole arch of the heavens for a time was wrapped in a sheet of flame, and the almost deafening crashes of thunder, following each other with scarcely an intermission between, seemed as if they would rend the solid earth, or topple it from its axis. A more sublime and awful meteoric display, I never witnessed or could conceive."[7]

The 30,000-odd Americans who trekked West in search of California gold in 1849 also witnessed their share of meteoric displays. It was a year of torrential rains and bitter cold on the Plains: one traveler noted in his diary that during the 43 days it took his party to travel from St. Louis to South Pass it rained 24 days; another wrote from the Plains that he had occasionally seen it rain just as hard back East "but neve [sic] seen it pour down by buckets full f[or] 6 hours incessantly."[8] Those expecting the parched wasteland of the Great American Desert were shocked to find themselves "wading to California," as one historian wrote. Others suffered for seemingly endless days in the hot, dry winds of the Great Basin, only to be overtaken by icestorms and snowstorms in the mountains.

The ill-fated Donner party was the most notorious weather disaster of the great mid-century westward migration. Delayed for a month in the desert during the summer of 1846, the eighty-nine pioneers, mostly members of the Donner and Reed families from Sangamon County, Illinois, did not reach the Sierra pass until the end of October and they were trapped by an early snowstorm. Storm after storm blew through from then on (the pass typically receives thirty-one feet of snow each winter), and by mid-December they were all but out of food. Eventually, the survivors resorted to cannibalism. When the rescue team finally reached the main group in February 1847, only forty-five people remained alive. "A more shocking scene cannot be imagined, than that witnessed by the party of men who went to the relief of the unfortunate emigrants in the California mountains," reported the *California Star* on April 10, 1847. "The bones of those who had died and been devoured by the miserable ones that still survived, were lying around their tents and cabins. Bodies of men, women, and children, with half the flesh torn from them, lay on every side. . . . Those who but one month before would have shuddered and sickened at the thought of eating human flesh . . . now looked upon the opportunity these acts afforded them of escaping the most dreadful deaths, as a providential interference in their behalf."[9] Weather and despair had together effected this horrifying transformation.

The Donner party remains a harrowing symbol of the perils of westward migration, but there were thousands of others whose deaths have been forgotten. It is estimated that as many as 10 percent of the pioneers died along the two-thousand-mile Oregon Trail—many of disease and accidents, but some, like the victims of the Donner party, of severe weather. The topography and weather of the West together created an immense buffer between the country's two halves. Topography and weather had to be tamed—if not in actuality, then at least in the public's imagination—before this buffer became part of America.

• • •

THE HISTORY OF HOW THE GREAT AMERICAN DESERT META-
morphosed, in the course of the nineteenth century, into the Garden of the
World has been recounted before—but it is a curious tale, and worth
repeating for the light it sheds on American weather attitudes. Until the
Civil War, American farmers had pretty much confined themselves to
regions where rainfall was reliable: the agricultural frontier had reached a
line roughly following the 96th meridian in eastern Kansas and Nebraska,
and then it made a giant leap over the Great Plains to Oregon and California. In between, as everyone knew, spread the Great American Desert, and
not even a fool would attempt to grow crops in a desert. But after the Civil
War, with editor and expansionist Horace Greeley's cry of "Go West,
young man!" resounding across the nation, settlements began to roll out
over the arid regions, creeping up the valleys of the Platte and Kansas
rivers. To the eyes of hopeful, land-hungry farmers and immigrants, the
vast, level, treeless plains in the west of Kansas and Nebraska and up in the
Dakotas began to look less like desert and more like farmland waiting to
be plowed. Now, as anyone knows who has lived or spent much time in
this part of the country, there is no fixed geographical barrier dividing the
lands that receive on average more than twenty inches of rain per year (the
bare minimum needed to raise crops without irrigation) from those that
receive less. It's not like driving up a mountain and passing in the space of
a few miles from grassland to conifer forest to alpine tundra. One perceives
the dryness of the High Plains more in the light and air and colors—"the
change is gradual," as Walter Prescott Webb writes in *The Great Plains,* "it
has about it the subtlety of a mood of nature."[10] But to Wallace Stegner,
who spent a good part of his life pondering aridity in the West, the change
is not gradual but "sharp." Here is how he describes the sensation of
crossing the 100th meridian in his biography of John Wesley Powell: "no
one need be in doubt about the sharp change in climate that occurs somewhere between the 96th and 100th meridians. It can be felt on the lips and
skin, observed in the characteristic plant and animal life, seen in the clarity
and/or dustiness of the atmosphere, determined by measurements of rainfall and evaporation, tested by attempts at unaided agriculture. Practically
every western traveler in the early years remarked the facts of aridity,

though not all used the word 'desert,' especially after the Kansas boom of the sixties had made the natives sensitive to supposed slurs."[11] Bryant was one of these early travelers: as his party crosses the meridian he notes that "the wagon-wheels have contracted so much from the effects of the dry atmosphere on the Platte, that the tires have become loose, and require resetting." Having complained of heavy dews and rainfalls heretofore, Bryant now gazes out in wonder on arid landscape: "not a cloud has exhibited itself on the face of the heavens, nor a tree or a shrub on the surface of the plain over which we have travelled, or in the distance as far as they could reach."

The "sharp change" that Stegner describes is a difference solely in *weather,* not in topography, for as far as the eye can see, the landforms are the same. If more rain fell, the difference between the eastern tall-grass plains and the western short-grass plains would disappear and the so-called desert would become garden. So Denver, with its fourteen inches of annual rainfall, and Laramie, Wyoming with its ten, and Salt Lake City with sixteen, and Helena, Montana, with ten could all become endless expanses of well-watered farms. If only more rain would fall. To the grandiose imagination of American expansionism, this was no obstacle. Indeed, those with an interest in seeing more rainfall on the plains yoked science (or pseudoscience) to imagination to prove that more rainfall was *already* falling in response to the spread of settlement. "Rain follows the plow" was the simple, catchy motto that crystallized the theory of climate change on the Plains, and it was really quite breathtaking in its simplicity and its arrogance. The theory crops up in various forms here and there in the arid states throughout the nineteenth century. In 1844, one Josiah Gregg, a Missourian who got his living in trading back and forth between Independence and Santa Fe, reports that the "old settlers" (those nameless old-timers who always seem to have the last word on weather history) told him that "droughts are becoming less oppressive in the West" and that "the rains" have much increased in New Mexico. Gregg attributes the "multiplication of showers" in the parched Southwest to the "genial influences of civilization" and the planting of "shady groves." A generation later, parties of eastern journalists trooped out West to have a look around and they told the same story of plentiful new rain soaking into the boundlessly fertile ground—and Horace Greeley made sure that every reader of his highly influential New York *Tribune* knew about it. Government-sanctioned scientific investigation came next: in 1867 Ferdinand V. Hayden, director of the Geological and Geographical Survey of the Territories, informed the Secretary of the Interior that climate change could be brought about in the arid West if each settler planted ten to fifteen acres of forest trees on each quarter section (160-acre plot). Other theories hatched out like new chicks: some attributed an increase in rainfall to the busting of the sod, as if puncturing the skin of grasses somehow released millennia of stored water vapor; some claimed the "vibration" of human occupation caused clouds to gather; others said that the trees that new settlers planted acted like magnets attracting rain clouds; or that explosions from guns or July 4 celebrations brought down the rains; still others pointed through the clouds to God himself, who was demonstrating his approval of the western settlements by watering their crops. "What of the 'Great American Desert,' which occupied so much space on the map a generation ago?" wonders one contemporary writer derisively. "It is *nomadic* and elusive; it recedes before advancing civilization like the Indian and buffalo which once roamed it . . ." The settlers had a terser way of saying the same thing: "The country is becoming more seasonable."

● "Rain follows the plow"—the slogan under which all these theories marched—was coined by Charles Dana Wilber, whom Henry Nash Smith identifies as "a speculative town builder and amateur scientist." Judging from his writing, Wilber was also a master of florid prose: "In the sweat of his face, toiling with his hands, man can persuade the heavens to yield their treasures of dew and rain upon the land." Man has the power to bring about "a grand consent, or, rather, concert of forces—the human energy of toil, the vital seed, and the polished raindrop that never fails to fall in answer to the imploring power or prayer of labor." Smith notes with his own rhetorical flourish that Wilber's slogan "makes the oldest and most sacred of agrarian symbols the instrument whose magical stroke calls down the life-giving waters upon the land." "Rain follows the plow": it has a commanding biblical ring to it, the righteous sound of revealed truth. But of course, this was not the first theory of climate change to roll across the face of America: far from it. As we've seen, the white settlers of the New World have been speculating about climate change ever since the first European ax bit into the soft white flesh of a North American tree. The "old settlers," and before there were old settlers, the old Indians, have been telling us right along that our presence on the land has changed the weather. Each new climate theory drew a different set of contrasts depending on the prevailing (or desired) weather of the day—wetter, warmer, more temperate—but the underlying process is the same in all of them: we the people have altered the climate by taking possession of the land.

What is remarkable about the "rain follows the plow" theory was how widely and vigorously it was promoted and how deep an impact it had. Politicians and railroad owners, town builders and land speculators eagerly took up the cry, sniffing profit and power to be gained by new settlers. "Why emigrate to Kansas?" queried a column in the *Western Trail,* a gazette put out by the Rock Island Railroad. "Because it rains here more than in any other place, and at just the right time." Naturally, when Horace Greeley took an "overland journey" from New York to California in 1859 and published his impressions in the New York *Tribune,* Kansas was "saturated with water." "Kansas brags on its thunder and lightning," he wrote in his dispatch from Lawrence on May 20, 1859; "and the boast is well founded. I never before observed a display of celestial pyrotechny so protracted, incessant and vivid as that of last Sunday night." During the boom years after the Civil War, the Union Pacific and Burlington railroads spent one million dollars just on promoting Nebraska, and a good portion of that budget went to broadcasting the "fact" of Nebraska's ideal climate.

The high priest of this campaign was William Gilpin, a sometime politician, entrepreneur, philosopher, soldier, lawyer, writer, and, as Marc Reisner adds, "windbag, and booby." Gilpin devoted a large chunk of his considerable energy to burying the Great American Desert, and he enjoyed the patronage of Missouri's powerful senator Thomas Hart Benton, the originator of the concept of Manifest Destiny. Together these two titans of expansionism endeavored to lure as many pioneers westward as they possibly could so as to fill what Gilpin called the "immense geography" of the

Plains. "While Benton sat in Missouri flogging pioneers westward," writes Reisner, "Gilpin stood in Colorado welcoming them and shrieking for more."[12] Gilpin went so far as to hatch an elaborate theory to support and validate his vision of the West. There was, he claimed, an "undulating belt" of about thirty degrees of latitude encircling the Northern Hemisphere: this geographic belt, in which the mean annual temperature is 52° F, contains an "axis of intensity" ideally suited to the flowering of higher cultures. "Within this isothermal belt," wrote Gilpin, "and restricted to it, the column of the human family, with whom abides the sacred and inspired fire of civilization, accompanying the sun, has marched from east to west, since the birth of time." Athens. Rome. Paris. London. New York. Topeka. Denver. The march westward was inevitable, preordained by God, climate, and geography. Theories linking climate and civilization have cropped up again—in fact, the concept is still alive and kicking today; but Gilpin's version was both the most outrageous and the most influential.

Incredibly enough, for a time it looked as if Gilpin and his collaborators were right. As pioneers streamed westward in the 1870s, they were blessed by a series of unusually wet years. The treeless plains yielded easily to the plow, and harvests were heavy. Gilpin, Greeley, and the railroad publicity machine made sure that word of rain verily following the plow spread far and wide. After 1878, the stream of settlers opened out into a flood—with half a meridian of Plains geography engulfed each year. Between 1870 and 1880, the population of the wheat states—Nebraska, Kansas, the Dakotas, and Minnesota—grew from less than a million to more than 2.5 million. Settlement was almost as rapid in the Pacific Northwest and the Rocky Mountain states. The great terrifying western buffer was being erased.

AND THEN THE RAIN STOPPED. AND BLIZZARDS HOWLED IN from the north. And clouds of grasshoppers descended and devoured the wheat. The pioneers who settled the West learned promptly enough that the "isothermal belt" was not as benign as Gilpin had promised. The journals, local histories, and stories written by the first generation of pioneers convey some of the shock—and disappointment—that these folks experienced once they had endured a few seasons west of the 100th meridian. One reads of 110-degree days in Kansas where the wind blows so hot that it crisps the crops in the field and farmers' sweat evaporates so fast that their dry faces become encrusted with the salt. The farmers in western Kansas joked that their summers were so brutally hot that the corn popped in their fields and that sinners had to be buried wearing overcoats so as to be comfortable in hell. The wind that blew almost constantly across the Plains drove many people out of their minds, particularly women. Dorothy Scarborough dramatizes this very phenomenon in her novel *The Wind*, in which a woman from Virginia succumbs to the maddening winds of the west Texas plains. Hamlin Garland, who moved as a young man from Iowa to South Dakota, roused tremendous controversy when he dared to tell the truth about the harsh conditions—and hard weather—of life on the Plains in his 1891 classic *Main-Travelled Roads*. In *A Son of the Middle Border*, his memoir of his boyhood and young manhood in the West, Garland describes the terrible ten-year drought that set in during the 1880s after rain ceased following the plow:

> An ominous change had crept over the plain. The winds were hot and dry and the grass, baked on the stem, had become as inflammable as hay. The birds were silent. The sky, absolutely cloudless, began to scare us with its light. The sun rose through the dusty air, sinister with flare of horizontal heat. The little gardens on the breaking withered, and many of the women began to complain bitterly of the loneliness, and lack of shade. The tiny cabins were like ovens at midday.[13]

By 1889, drought had burnt up the few tokens of softness and gentility that his parents had tried to transplant out to the desolate Plains: "The trees which my father had planted, the flowers which my mother so faithfully watered, had withered in the heat. The lawn was burned brown. No green thing was in sight, and no shade offered save that made by the little cabin. On every side stretched scanty yellowing fields of grain, and from every worn road, dust rose like smoke from crevices, giving upon deep-hidden subterranean fires." "The disaster is never sudden, but drawn out over days, weeks, and months," writes Webb of drought on the Plains. "The suffering is no less terrible because it is fraught with the persistent hope of rain. . . . One can flee from a flood or a storm, but one does not flee from a drought. In too many cases by the time hope is lost, the means of fleeing have departed."[14] As one settler concluded after surveying the effect of prolonged drought on the land, "There is no god west of Salina."

Drought inflicted a slow desiccating death on the dreams inflamed by Gilpin's rhetoric. But in the frigid winters, blizzards killed instantly and without warning. The very word "blizzard" may be a coinage of the western Plains, though its etymology has been hotly disputed. After the famous East Coast blizzard of March 12, 1888, the New York *Times* noted primly that the word was "simply a bit of onomatopoeia . . . supposed to sound more or less like the thing it denotes." The English, however, claim it comes from an expression common in the Midlands—"may I be blizzered," which means something like "I'll be damned," the idea being that the speaker is knocked down as if by a blizzard's icy blast. The Germans, on the other hand, insist the word comes from *blitzartig,* meaning "lightning-like." According to Plains historian Webb, "blizzard" was first employed in a meteorological sense by O. C. Bates, "the erratic editor of an Iowa newspaper," when he needed a noun to match the ferocity of the snow-storm that howled through on March 14, 1870. The Oxford English Dictionary reports that "the word became general in the American newspapers during the severe winter of 1880–81." Blizzards—which to qualify officially today must have temperatures of 20° F or below and wind speeds of at least 35 miles per hour—do occasionally hit the East, but they are far more common, and usually far more severe in the West, especially on the Plains (the East Coast blizzard of 1888 was described proudly in the New York press as a "real Western-style blizzard").

With no natural barriers of any consequence between the Gulf of Mexico and Canada ("The only thing between us and the North Pole is a bob-wire fence," the farmers like to say, "and that blew down two years ago"), winter storms whip down from the northwest with blinding speed and awful intensity. "The blizzard was a storm peculiar only to the open plains," writes Everett Dick in his early social history *The Sod-House Frontier, 1854–1890.* "It was less of a snow storm than an ice-dust wind storm which drove a smother of pulverized ice into the air from the ground and carried it along in a veritable cloud of icy particles which beat with such stinging cold that neither man nor beast could stand to face it. The whole landscape was obscured by this sweeping frigid storm."[15] Farmers commonly ran ropes or clotheslines between house and barn so as not to lose their way in the impenetrable blasts of snow—but even so many perished in the storms. One reads of people caught out in blizzards on the Plains who froze to death just steps away from their homes—the blowing snow rendered them blind and their cries for help were lost in the deafening wind. The powdery snow, as fine as flour, "found its way through every crack and crevice," one nineteenth-century North Dakotan wrote, and settlers often woke up on blizzard mornings under a frozen white blanket four inches deep. Dick writes memorably of the "schoolchildren's storm" of 1888—a blizzard that blew up so fast that children in Nebraska and the Dakota Territory were overtaken on their way home from school: "The morning of January twelfth dawned clear, warm, and bright. The children started cheerfully off to school. Before long a fifty-five-mile wind was blowing and the temperature dropped to thirty-six degrees below zero in Nebraska. Children coming home from school were caught. Parents coming after their children lost their way and were frozen."[16] It was the speed and unpredictability of blizzards that were most terrifying and often most dangerous.

One South Dakota survivor of the '88 blizzard remembered playing outside at recess in shirtsleeves and suddenly glancing up at "something coming rolling toward us with great fury from the northwest, and making a loud noise. It looked like a long string of big bales of cotton, each one bound tightly with heavy cords of silver, and then all tied together with great silvery ropes." When the children alerted their teacher to the approaching clouds, she rushed out to ring the school bell, and the children "made a mad rush for the schoolhouse. The last one just got inside the building when the storm struck. It hit the building with such force that it nearly moved it off its cobble-stone foundation."[17] Even on calm, cloudless days, settlers took to scanning the northwestern horizon for the smudge of black that betokened a storm.

Blizzards were not the only violent weather on the plains and prairies. Cold fronts sometimes raced down from Canada with unbelievable swiftness, plunging temperatures 50 degrees or more in a matter of minutes. A settler named John Moses recorded one such front that hit Illinois on December 21, 1836: "The water in the little ponds in the roads froze in waves, sharp edged and pointed, as the gale had blown it. The chickens, pigs and other small animals were frozen in their tracks. Wagon wheels ceased to roll, froze to the ground. Men . . . caught out on horseback were frozen to their saddles, and had to be lifted off and carried to the fire to be thawed apart. Two young men were frozen to death near Rushville. One of them was found with his back against a tree, with his horse's bridle over his arm and his horse frozen in front of him."[18] Such outbreaks became part of the mythology of the region: the stories were told over and over and then passed down like heirlooms through the generations.

Many, many stories were told about the hard winter of 1880–81, a season of unprecedented severity and universal suffering on the Plains. Anyone who has read Laura Ingalls Wilder's children's classic *The Long Winter* (originally published in 1940) is familiar with this winter: the first blizzard arriving in October, and then storms hitting weekly from January through April, with each blizzard lasting three days or more; families forced to grind their seed wheat endlessly in coffee mills and burn twisted hay for fuel; train service completely interrupted for over ten weeks; store shelves picked clean as starvation looms ever nearer; snow so deep that it completely buried houses and raised town streets to the level of second-floor windows. "That blizzard seemed never to end," writes Wilder of the family's mood by midwinter. "It paused sometimes, only to roar again quickly and more furiously out of the Northwest. Three days and nights of yelling shrill winds and roaring fury beat at the dark, cold house and ceaselessly scoured it with ice-sand. Then the sun shone out, from morning till noon perhaps, and the dark anger of winds and icy snow came again."[19]

Children today read Wilder's "little house on the prairie" books as wonderful, cozy stories of the good old days—but everything she wrote about the weather was, in fact, true. More than eleven feet of snow really did fall in the Dakota Territory that winter, with no thaw to melt it. No trains ran for seventy-nine straight days. When the snow finally melted, the flat, flooded countryside was littered with the corpses of livestock that had perished in the storms. Human bodies were found in the spring as well —an appalling fact dramatized in another memorable novel of the Plains, O. E. Rölvaag's *Giants in the Earth* (originally published in 1924) about the first Norwegian settlers in the Dakotas. Rölvaag writes movingly of how one woman is driven insane by the desolation of the land and the bleak monotony of the winter weather. He describes the shock the immigrants experience when the first Plains blizzard overtakes them: back in Norway they "had seen plenty of storms that made up fast; but nothing like this had ever before come within the range of their experience. Like lightning a giant troll had risen up in the west, ripped open his great sack of woolly fleece, and emptied the whole contents of it above their heads."[20] Only with a biblical metaphor can Rölvaag convey the suffering the people experienced during the hard winter of 1880–81: "They say it rained forty days and forty nights once in the old days, and that was terrible; but during the winter of 1880–81 it snowed twice forty days; that was more terrible. . . . Day and night the snow fell. From the 15th of October, when it began, until after the middle of April, it seldom ceased. From the four corners of the earth it flew; but of all the winds that brought it, the south wind was the worst."[21] Wilder's *The Long Winter* ends happily with the Ingalls family celebrating Christmas in May when the train finally comes through; but *Giants in the Earth* is a tragedy: the powerful Hans Olsa is stricken with a deathly illness after being caught out one night in a blizzard while tending to his cattle, and the novel's hero, Per Hansa, freezes to death on the open plain while skiing off to get the minister for his dying friend. He is discovered only the following May, sitting on a haystack with his glazed eyes set toward the west.

Five years later in 1885–86 another savage winter hit the Plains, slaying as many as 85 percent of the cattle in some regions. The "cattle kingdom" of American legend, with its colorful roundups and cattle drives and tuneful cowboys, collapsed as a result of those winter storms. It was a short-lived bonanza, lasting "a scant two decades," as one historian writes, before it was "bankrupted by weather and, more, by overexpansion."

Still another meteorological killer that concentrates its fury on the Plains states is the tornado. Three-quarters of the world's reported tornadoes hit the United States, and the majority of these form over the Great Plains. In the "tornado alley" that runs from north-central Texas through central Oklahoma, Missouri, Kansas, Nebraska, and north into the Dakotas, an average of 700 tornadoes strike each year: they come for the most part in the late spring and early summer, spawned when warm humid air from the Gulf of Mexico is forced to rise suddenly over the denser sinking air of an advancing cold front. Tornado winds, which rotate counterclockwise in the Northern Hemisphere, have topped 300 miles per hour; they are notoriously difficult to predict and erratic in their movements, size, and duration, with life spans ranging from a few seconds to seven hours and diameters varying from a few feet to two miles across. The apparition of the funnel cloud spiraling earthward out of a blackened sky has seared itself onto the imagination of the West. As John P. Finley of the U.S. Army Signal Corps (forerunner of the National Weather Service) wrote in his 1887 book on the subject, "A single experience of this awful convulsion of the elements suffices to fasten the memory of its occurrence upon the mind with such a dreadful force that no effort can efface the remembrance of it."[22] The early settlers of the West who did experience such convulsions amply confirm this claim. Finley's book includes this eyewitness account of the pair of tornadoes that hit the town of Irving in Marshall County, Kansas, on May 30, 1879:

In the wake of the first tornado a warm southerly wind passed over the town accompanied by rain. The sun, now partially exposed beneath the heavy clouds lining the western horizon, threw its warm rays upon the terror-stricken inhabitants, who, at this welcome invitation, assuring them as they thought of peace and protection, emerged from their cellars and dug-outs to witness the destruction already committed and relieve their suffering neighbors. Hardly had the people recovered from the first shock, when there appeared in the West a cloud of inky blackness and enormous dimensions, presenting a square front of apparently two miles in width and a perpendicular height from earth to sky. It moved along slowly, but with the most inconceivable majesty of force, annihilating everything within its reach. The cloud is now at the outskirts of the town, and as it begins to execute its frightful mission of death and destruction the earth fairly quakes and trembles. All nature stands aghast, and every living thing seeks, but in vain, to find security from the impending danger. Many people actually believe that the Judgment Day has come, and offer fervent prayers and loud appeals for preservation. But the hand of mercy stays not the dreadful carnage. It begins. The awful roar, like the belching forth of a thousand Columbiads, drowns the most piercing cries of the wounded. The cloud strikes into a cluster of eighteen houses and other buildings filled with human beings and the accumulations of years.[23]

Like a demon, the storm kills with perverse ingenuity and seemingly intentional malice: families are blown hundreds of yards into fields and mangled to death; a twenty-year-old woman is stripped of all her clothing, her flesh lacerated and caked with mud, almost every bone in her body is broken. Finley draws the grim moral for his readers: "This prosperous community had been scourged by a fell destroyer more dreadful than either flood or fire, epidemic or war. It came in the twinkling of an eye and all was gone; life, property, happiness crushed and annihilated; swept with lightning speed into eternity."

Perverse killers, tornadoes have on occasion shown a perverse kind of mercy. One thinks of the "cyclone" that transported Dorothy from dull, gray Kansas to the magical land of Oz in L. Frank Baum's 1900 classic *The Wonderful Wizard of Oz*. Dorothy's unharmed ascent through the still center of the funnel cloud may not be as fantastic as it appears. Consider this bizarre but factual bit of tornado lore from a May 1984 article on severe weather that William Hauptman published in *The Atlantic*:

A few very fortunate people have gone aloft in tornadoes and survived. During [a] tornado in Wichita Falls, a man was blown out of his exploding house. Like Dorothy, he glimpsed others in the funnel. A house trailer rotated near him, and in the window he could see the terrified face of one of his neighbors. (She did not survive.) Flying ahead of him was a mattress. If I could reach that, he thought, I'd just go to sleep. He then lost consciousness and woke on the ground, wrapped in barbed wire. Flying splinters had made a pincushion of his body.

"Man have power?" demands Rölvaag in *Giants in the Earth* in the face of the ferocious storms of the Dakotas. "Breathe it not, for that is to tempt the Almighty!" In the great mythology of the settling of the American West, man and weather are adversaries, locked in a death struggle for dominion over the huge, mysterious land.

A CURIOUS AND RECURRING EPISODE IN THAT LONG STRUGGLE has been man's attempt to control or modify the weather of the West, especially its scant rainfall. The agricultural Native American tribes of the Plains and the arid Southwest, as we've seen, had a whole slew of weather-influencing myths and rituals, and the white settlers began to add their own as soon as rain ceased following the plow. "Pluviculture" became a burgeoning field in the arid West, with rainmakers of various stripes promising quick results from schemes like setting immense fires, shooting cannons at clouds, or releasing balloons loaded with explosives. This last technique was the one favored by R. G. Dyrenforth, whom the Secretary of Agriculture hired in the 1890s to conduct rainmaking experiments in Texas. Dyrenforth's decidedly mixed results earned him the nickname "Dryhenceforth."

Charles M. Hatfield was perhaps the most famous—and infamous—of America's "cloud compellers": operating mostly in southern California in the early years of the twentieth century, he developed an original technique of attracting storm clouds by placing pans full of his secret (and supposedly quite pungent) chemical concoction on platforms at the top of wooden towers. "These gases smell so bad that it rains in self-defense," is how a San Francisco journalist explained the process.[24] Strangely enough, it seemed to work. Hatfield's career took off when he won a much-publicized $1,000 bet that he could bring at least eighteen inches of rain to Los Angeles between mid-December 1904 and late April 1905; and he claimed other big successes that year in northern California and the San Joaquin Valley. The Weather Bureau launched an all-out campaign to discredit him, insisting that they had predicted every one of these rainstorms; but Hatfield's reputation remained untarnished as long as rain coincided with his rank ministrations—which it did with uncanny regularity year after year. The most notorious episode of his career was "Hatfield's Flood": he had promised to end a prolonged drought in Moreno, about sixty miles east of San Diego; but he must have brewed up an especially potent batch of rainmaker, for the record precipitation of January, 1916 not only filled the reservoir but also broke through a dam, resulting in a flood that killed an estimated fifty people and brought extensive damage to San Diego.

Later in the twentieth century, explosives and chemical soups were set aside in favor of cloud seeding—sprinkling dry ice or silver iodide crystals into clouds to make them release their moisture content. By the 1940s, pluviculture had shed much of its quackery and gained a fair measure of respectability. The government, at both the state and federal level, has at various times sponsored rainmaking experiments in the dry western states; and hopes run high every generation or so that we may at last be on the verge of banishing drought altogether. But so far, substantial and practical weather modification has eluded both scientist and charlatan.

"Man have power? Breathe it not."

A SPELL OF FAVORABLE WEATHER AND THE PROMISE OF MORE to come spurred the settlement of the Plains states. When the blizzards and droughts of the 1880s proved these promises cruelly false, hundreds of thousands of homesteaders gave up hope and abandoned their claims. By

1890, the third straight drought year, Kansas had lost one-quarter of its population, Nebraska half. According to Marc Reisner, of the more than a million families who had tried to make a go of homesteading on the Plains, only 400,000 remained after the drought years of the late nineteenth century. "Weather was the ultimate arbiter in the American West," writes Reisner. "Unless there was some way to control it, or at least minimize its effects, a good third of the nation might remain uninhabitable forever." This should not have been news even in the nineteenth century. More than a decade before the weather decimated the farmers of the Plains states, the Government Printing Office had issued John Wesley Powell's exhaustive and authoritative *Report on the Lands of the Arid Region of the United States.* Powell, who led the first heroic team of explorers down the Colorado River in 1869 and made additional exploring forays into the West during the 1870s, was acutely interested in the interaction of climate and topography. He had seen with his own eyes the essential environmental fact that Gilpin and his supporters had devoted their careers to denying: that much of the West was just too dry to farm. "Westward the amount of aqueous precipitation diminishes in a general way," Powell states bluntly on the first page of his *Report,* "until at last a region is reached where the climate is so arid that agriculture is not successful without irrigation. The Arid Region begins about midway in the Great Plains and extends across the Rocky Mountains to the Pacific Ocean."[25] As Wallace Stegner points out, Powell's *Report* forcefully challenged "the folklore bred up through generations of frontier farmers in a country of plentiful rain." Powell does pause to acknowledge the noisy claims of the "rain follows the plow" school and he considers in turn each of the "many conjectures and hypotheses" explaining this phenomenon. But logic and science lead him to reject them all out of hand. "If it be true that increase of the water supply is due to increase in precipitation, as many have supposed," he writes cautiously, "the fact is not cheering to the agriculturalist of the Arid Region. The permanent changes of nature are secular [lasting through many ages]; any great sudden change is ephemeral, and usually such changes go in cycles, and the opposite or compensating conditions may reasonably be anticipated. . . . If [the increase in the water supply] is due to a temporary increase of rainfall . . . we shall have to expect a speedy return to extreme aridity, in which case a large portion of the agricultural industries of the country now growing up would be destroyed."[26] "The land is not one of steady deficiency that men can count on, as is the true desert," writes a modern historian of the region, "but one of sharp extremes—heat and cold, floods and droughts, cyclones and blizzards. It is an unreliable, intractable place, wildly oscillating around an almost meaningless mean. The only certainty is that droughts will come, and come often, and that in 'normal' years the region lies at or below the rainfall margin for most farming."[27]

Droughts came and destroyed the "agricultural industries" in the 1880s. And destroyed them again in the dust storms of the 1930s and again in the 1950s. The drought of the 1930s is the one everyone remembers because it was the longest and the most destructive, and because it hit during the dark days of the Great Depression. Once again, human ignorance and delusion played into the hands of nature's fickleness. Historians trace the 1930s dust bowl back to a huge increase in the demand for wheat that occurred during World War I and that kept prices high into the 1920s. As long as prices stayed high and the rains kept falling—which they did with eerie abundance and regularity year after year from the teens through the 1920s—farmers kept on plowing and planting. During the "great plow-up" of 1925–30, 5,260,000 acres of the southern Plains were stripped of their natural vegetation. "When the black blizzards began to roll across the region in 1935," writes western historian Donald Worster, "one-third of the Dust Bowl region—33 million acres—lay naked, ungrassed, and vulnerable to the winds." That region encompassed western Kansas, southeastern Colorado, the Oklahoma panhandle, the Texas panhandle, and northeastern New Mexico. The term "dust bowl," like the meteorological use of "blizzard," is attributed to a western newspaper reporter, in this case Robert Geiger, an Associated Press stringer on assignment in the southern Plains, who was trying to find language for the human suffering he saw around him after the notorious Black Sunday of

April 14, 1935: "Three little words, achingly familiar on a western farmer's tongue, rule life today in the dust bowl of the continent—if it rains."

"Roller" or "black blizzard" was the Plains term for the worst dust storms: a roller was an immense cloud of topsoil that billowed four miles into the atmosphere, blotting out the sun's light, flaying off the skin of anyone caught out in it, filling lungs, sifting into houses no matter how painstakingly they were caulked. "Wind forty miles an hour and hot as hell," a Nebraskan wrote in his diary in the summer of 1934. "Two Kansas farms go by every minute." According to more precise, if less colorful, estimates, more than four and a half tons *per acre* entered the atmosphere during the five black blizzards of the spring of 1935. In his book about the dust bowl of the southern Plains, Worster vividly describes how a black blizzard arrived "with a rolling turbulence, rising like a long wall of muddy water as high as 7000 or 8000 feet. Like the winter blizzards to which they were compared, these dusters were caused by the arrival of a polar continental air mass, and the atmospheric electricity it generated helped lift the dirt higher and higher in a cold boil, sometimes accompanied by thunder and lightning, other times by an eerie silence."[28] It was the Black Sunday roller that inspired Woody Guthrie's famous ballad "So Long, It's Been Good to Know You." On that same day the Reverend Rolley Wells of Guymon, Oklahoma, led his congregation in a special "rain service": but unlike the Pilgrim fathers who prayed their way out of a Massachusetts drought in 1623, the Methodists of Guymon were disappointed, for the dust blew in that very afternoon, just after they returned home from church. "The story of the southern plains in the 1930s is essentially about dust storms, when the earth ran amok," Worster concludes. "And not once or twice, but over and over for the better part of a decade: day after day, year after year, of sand rattling against the window, of fine powder caking one's lips, of springtime turned to despair, of poverty eating into self-confidence. . . . [T]he dust storms that swept across the southern plains in the 1930s created the most severe environmental catastrophe in the entire history of the white man on this continent."[29] Environmental catastrophe also, inevitably, became human catastrophe.

Worster notes that 3.5 million Plains farmers left their homes in the course of the 1930s. "Not all were dusted out," he writes "but all were uprooted —a generation of human tumbleweeds, cut loose from the soil."[30]

To residents of Chicago, Detroit, Washington, and New York, the dust was not just an abstraction—something they read about in newspapers or heard of from relatives. The worst of the blizzards darkened their skies as well. A May 1934 roller dumped 12 million tons of Plains topsoil on Chicago. Noon was as dim as evening the next day in Buffalo and the following day Savannah reported a dusty haze in the air. The blowing dust made America's whole western horizon boil black, as if the very continent were on fire. But fire would have been mild compared with the devastation inflicted by the dust storms. The dust bowl, a tragedy for both the land and the people who lived on it, was the direct consequence of a century of failure to reckon with the climate of the West.

THAT FAILURE GOES RIGHT ON TODAY. AS BERNARD DE VOTO has written, Americans have never ceased "believing that there must be some way of licking climate or that climate would adapt itself to men's desires." Out West, hope blooms anew each year, like the fleeting spring green of the Plains grasses. Back in the 1960s, a sociologist named Thomas Frederick Saarinen trekked around the arid counties of Colorado, Oklahoma, Kansas, and Nebraska, interviewing local farmers about their beliefs and attitudes concerning drought. He showed them photographs of drought-stricken farms and recorded their comments, he compared their recollections about past droughts with climatological statistics, he got them to fill out questionnaires about the productivity of their farms and their dependence on rainfall. Collating all his data, Saarinen concluded that farmers are obsessed with rain, talking about it "frequently, animatedly, and in dry spells, almost exclusively"; and yet even though all of the farmers in his sample had suffered through severe droughts, most of them remained "eternally optimistic" about their region's climate. They tended to remember their first and worst droughts and to forget all the others, except the most recent ones. Many cited the Bible to explain their rough, unpredictable weather—seven fat years and seven lean years; others blamed drought on the government's testing of atomic bombs (this was before the test-ban treaty); still others relied on inherited folk wisdom. However they accounted for the vagaries of their scant rainfall, they managed to convince themselves that next year would be better. As one wheat farmer said, when asked to comment on a grim photo of a man and two boys walking with their heads bowed into a dust storm on a desolate landscape: "The man and the older boy they're going to get in out of it anyway. He's going to tell those boys when they get in there, 'Well we didn't raise any crops this year but wait 'til next year.' Because that is the biggest next year country in the world out there."[31] As Webb has written, "Whatever man desires he attains—if not in reality, then vicariously through his imagination. Men in the Great Plains, or the arid region, have wanted rain more than they have wanted anything else."[32]

The biggest next-year country in the world. This motto holds true not only for farmers on the arid Plains but also for thousands who swarmed into California during the gold rush and the millions more who emigrated to the Golden State during the twentieth century in search of jobs, glamour, the good life—and of course, always, year-round sunshine. In California good weather—warm, sunny, comfortable, relaxing, and, yes, perpetually dry weather—has engendered an entire culture revolving around the daily appearance of the sun. The weather is not just beautiful in California, it's the very fount of health, long life, easy lifestyle. "The air is so uniformly dry and bracing," marveled Horace Greeley during his 1859 sojourn in the state, "and the days so generally tempered by a fresh breeze, that the human frame maintains its elasticity in spite of severe and continued exertion."[33] Marion Thrasher, a San Francisco doctor, elaborated on the same point in his remarkable little book of 1915, *Long Life in California:* "It seems that a miracle has been wrought by the forces of nature in the perfection of our California climate," he writes. "The warm genial climate of the Pacific Coast, where life runs easy, and money [is] made without effort, where robust health is stamped on every brow," assures Californians of "an unparalleled longevity."[34] And indeed, to prove his point, Dr. Thrasher compares California's average life expectancy of 74 years with that of damper, colder places—47 in Massachusetts, 37 in England, 20 in India. The human race perfected through perfect weather. Thrasher makes no mention of the darker side of this perfect weather—the prolonged summer droughts, the fickle winter rains, the monotony of day after day of the same weather. But Greeley, western booster though he was, did register his alarm at what happens to the landscape after a season without rain: "The traveler who first looks down on the valleys and lower hillsides of California in midsummer is generally disappointed by the all but universal deadness. Some hardy weeds, a little sour, coarse grass along the few still living watercourses, some small, far-between gardens and orchards rendered green and thrifty by irrigation, form striking exceptions to the general paralysis of all the less inspiring manifestations of vegetable life." Contemplating the prospect of relentless sun from June to November, Greeley confesses that "my early prejudices in favor of a refreshing shower occasionally are not fully overcome. I dislike to look for miles across so rich and beautiful a valley as this of San Jose, and see paralysis and death the rule, greenness and life the exception."[35]

Greenness and life did, in time, spread throughout the dead brown valleys of California, when massive irrigation projects redirected rivers, sluiced down melted mountain snows, and pumped out aquifers. Water made the summer deserts into astonishingly productive farmland; water allowed the city of Los Angeles to grow from a dusty Spanish mission town to the far-flung metropolis it is today. But it was borrowed or in some cases stolen water, seized through epic engineering endeavors that were propelled by greed, corruption, deceit, and towering ambition. Roman Polanski dramatized one particularly vicious chapter of California's water wars in his powerful film *Chinatown*. Marc Reisner documents the complete story in his book *Cadillac Desert: The American West and Its Disappearing Water*. "California, which fools visitors into believing it is 'lush,' " writes Reisner, "is a beautiful fraud. . . . The whole state thrives, even survives, by moving water from where it is, and presumably isn't needed, to where it isn't, and presumably is needed. No other state has done as much to fructify its deserts, make over its flora and fauna, and rearrange the hydrology God gave it. No other place has put as many people where they probably have no business being. There is no place like it anywhere on earth."[36]

California's extremity may be unique, but the "beautiful fraud" of weather and landscape has infected all of the West in one way or another. "There is something about this region that seems to intensify the tendency to live in one world while dreaming of or expecting another," writes Worster. "The West is characteristically a country of daydreams and fantasies, of visions and nostalgia." And of course in the background of every dream are the limitless blue skies and immense clear distances of the West. Listen to how Rölvaag opens his epic novel *Giants in the Earth:* "Bright, clear sky over a plain so wide that the rim of the heavens cut down on it around the entire horizon. Bright, clear sky, to-day, to-morrow, and for all time to come . . . And sun! And still more sun! It set the heavens afire every morning; it grew with the day to quivering golden light—then softened into all the shades of red and purple as evening fell . . ." Or read one of John Muir's rhapsodies about the crystal clarity of the atmosphere of Yosemite: "Probably more free sunshine falls on this majestic range than on any other in the world I've ever seen or heard of. It has the brightest weather, brightest glacier-polished rocks, the greatest abundance of irised spray from its glorious waterfalls . . . more starshine, moonshine, and perhaps more crystal-shine than any other mountain chain, and its countless mirror lakes, having more light poured into them, glow and spangle most . . . How fine the weather is! Nothing more celestial can I conceive. How gently the winds blow! Scarce can these tranquil air-currents be called winds. They seem the very breath of Nature, whispering peace to every living thing."[37]

Free sunshine. Fine weather. Peace to every living thing. One could almost set the words to music. But look a little closer, and you'll see that the very hymns of praise are wrapped around a black hole of paradox. The essence of the West is the freedom born of "bright, clear sky" and vast open space—space that is itself "the product of incorrigible aridity," as Stegner puts it. So the heaping plenty of the West is inextricably bound up with its scarcity, abundance with absence, grandeur with negation. Worster plays with the same idea: "Freedom in our western vision requires aridity. It depends on a brilliance of light, an openness of terrain, a clean spaciousness that gives us plenty of room to spread out and look around. . . . It requires the West as it naturally was and is. A little more water might spoil it." The freedom of aridity is ultimately a mirage, a freedom that will someday dry up or run through our fingers. A land where the skies are not cloudy all day does not get enough precipitation to sustain a civilization of irrigated crops, swimming pools, air-conditioned houses and cars, long hot showers, and daily-sprinkled lawns. The great bulk of this civilization has been hurriedly thrown up in the past half century—but recent evidence from tree-ring analysis reveals that the period from 1937 to 1986 was abnormally wet in the western states.[38] The prolonged drought that hit the West, and especially California, after 1986 was not a freak occurrence, but a natural and expected part of the climatic cycle. The West, to borrow Webb's term, is an oasis civilization, and history tells us that every oasis civilization has sooner or later shriveled up, salinized its lands, and died. Assyria, Carthage, the Aztecs, the Hohokam Indians of the Southwest: all gone without a trace.

We know this. Our most distinguished western writers and environmentalists keep explaining it to us. The statistics look more alarming every year. And yet we continue to live inside the illusion of freedom. The immense Ogallala aquifer, an underground water supply that roughly coincides with the dust bowl region, was for years assumed to be infinite; pumping accelerated like a rocket after World War II, with the total number of acres under irrigation jumping from 3.5 million in 1950 to 15 million by 1989. But now in Texas and New Mexico wells tapping down to Ogallala waters are starting to run dry. It's only a matter of time before the same thing starts happening in other regions as well. There just isn't enough natural precipitation west of the Mississippi for us to go on living the way we like to live.

It's pushing two centuries now since Lewis and Clark journeyed from St. Louis, Missouri, to the Pacific Ocean, and still we don't quite know what to make of the weather of this region. Its wild swings from one extreme to another, its frigid, blizzard-ridden winters and parched wildfire summers; its tornadoes and flash floods; its long smiling spells of gorgeous serenity that come crashing to an end in vicious squalls; and above all its inescapable aridity—all of this continues to astound us. Rain followed the plow for a while, but then it stopped. The tremendous network of federal and state irrigation projects, after all these years, has only greened an area the size of Missouri. But we still don't get it. We just can't wrap our brains around the fact that more than half of the acreage of the continental United States is classified as deficient in moisture. "A semidesert with a desert heart," Webb called it. The United States—a predominantly arid or semiarid country? Impossible. Sunny weather—yes. Semidesert—no. And so we bask under flawless skies and turn on the tap full blast. A beautiful fraud.

5.

FROM PROVERB TO PROBABILITY: WEATHER FORECASTING IN AMERICA

"Can any understand the spreadings of the clouds?"

—Job 37:16

WEATHER FORECASTING, LIKE FORTUNE-TELLING, APPEALS TO something so deep in us it practically qualifies as an instinct. Even if tomorrow's weather is going to be pretty much like today's, we want to know about it. Getting a weather forecast gives us a feeling of participating in nature's mystery, a sense of control over the uncontrollable, a hedge against disaster—or at the very least something to complain about. And the farther out in time the forecast goes, the more fascinating it becomes. The fathomless wisdom and daring of someone who stands before us and shows us the next five days, the coming month, the season. If only we could believe it! Part art, part science, weather forecasting even in this age of supercomputers has not shed its aura of magic. We scoff, we rail, we dismiss it as deliberate lies, hocus-pocus, a bad joke, an assault on our common sense—but we fall for it every time. There's always that possibility that "they" will be right this time, or next time, or the day after. Weather forecasting is a powerful business because the real power comes from our desire to know.

Here in America, as settlement spread westward in the course of the nineteenth century, the power of forecasting took on a special urgency. As more and more territory got "settled up," the weather itself expanded. Entire weather systems fell within the borders of the United States. What was happening on the West Coast was seen to have a direct impact on the weather of the Rockies, and then the Plains, and eventually the Midwest and East. The very size of the country became a major catalyst in pushing forward the science of meteorology during the nineteenth century. Those old folk sayings about solar halos and bird flight might work for the farm, the town, the county. But traditional weather lore could not be stretched across 3,000 miles of varied terrain. The expanding country *needed* a more rational, systematic approach to forecasting. The fact that weather science came of age simultaneously with the nation's spanning of the continent is surely no coincidence. The two events went hand in hand, bound together by the development of the telegraph and the evolution of a federal weather service.

Of all the various forecasting tricks and traditions that Europeans brought with them to America, the most curious and the most strangely enduring is the almanac. "Second in popularity only to the Bible," according to one historian, almanacs containing yearlong forecasts and rules for weather prognostication began appearing in Germany in the early sixteenth century and quickly spread through the continent and into England. Almanac writers based their long-range forecasts primarily on astrology, with a good measure of the Christian calendar, hocus-pocus, and of course pure serendipity tossed in. Thus, the weather on Christmas Eve and Christmas Night was supposed to determine the entire year's cycles of wet and dry, hot and cold, unless such influence was overruled by the alignment of Scorpio and Capricorn. It's impossible to say whether almanac writers really *believed* their yearly outlooks, but it really doesn't matter. The almanacs were like today's supermarket tabloids: they forged their own reality. They hit a nerve that has never stopped tingling.

Ben Franklin, who was both weather wise and business wise, tapped into the lucrative American market with his famous *Poor Richard's Almanac,* first published in 1732 and a best-seller for the next quarter century, with sales averaging 10,000 copies per edition. Franklin the scientist was well aware that astrological forecasting was bunk, and he ridiculed the practice in every edition of *Poor Richard's.* But Franklin the businessman knew that yearly forecasts sold books, and so, with a wink and a smile, the whimsical prognostications went out year after year. A Franklin forecast was never too precise and always highly amusing. Here's a sampling from *Poor Richard's, 1733:* January 7: "snow if not too warm about this time"; March 19–20: "March many weathers"; April 12–20: "Wind or Thunder and Rain. Beware of meat twice boil'd and an old foe reconcil'd"; July 19: "Dog days begin"; "August 21: "Dog days end" and so on.

In 1792 *The Farmer's Almanac* picked up where Franklin left off, and it's still going strong today as *The Old Farmer's Almanac,* with a year's worth of weather interspersed with household advice, bits of history, lunar and solar charts, recipes, ephemera, whimsy, and anecdotes. A recent history of the first two hundred years of *The Old Farmer's Almanac* reveals that the "secret weather forecasting formula" of almanac founder Robert B. Thomas "still exists and is kept safely tucked away in a black tin box at our Almanac offices in Dublin, New Hampshire. Basically, it's a rather complex series of weather cycles Thomas and others of his time observed." Rank nonsense, we might mutter to ourselves, but who can resist taking a peek now and then into this folksy weather crystal ball? There is something so lulling in its confidence, so seductive in its air of mystic arcana. "Weather Forecasts, Planting Tables, Zodiac Secrets" the cover promises. *The Old Farmer's Almanac* has managed to assume the unassailable authority of an old farmer who's been watching the skies for so long that he has *become* the weather: Ignore me at your peril, for I have the wisdom of the ages flowing through my veins. And, to add a new wrinkle, some contemporary researchers claim that "astrometeorology" may have some rational underpinning after all: they speculate that the shifting alignments of the planets could alter the center of mass in the solar system and thus exert an indirect impact on our weather. As we all know, almanacs *do* sometimes hit the nail uncannily on the head. So who's to say that the black box is pure hokum?

Folk sayings and proverbs—the "red sky at night, sailor's delight" school of forecasting—are in a different league altogether from the almanacs, for they are grounded in one of the essential principles of meteorology: close observation of nature over extended periods of time. For millennia before the arrival of whites, Native Americans forecast the weather on the basis of the behavior of animals and plants, the direction of the wind, the color of the sky at certain times of day, and atmospheric phenomena such as rainbows and halos. The European settlers brought their own weather folk wisdom, which they adapted to the climates of North America and grafted onto the Indian lore. The daisy shuts its eye before rain. Spiders enlarge and repair their webs before bad weather. When dogs eat grass rain follows. A solar halo indicates bad weather. Much rain in October, much wind in December. There are thousands of these "prognostics," varying by region, covering every conceivable type of weather. Some are patently absurd, especially the sayings concerned with long-term forecasts: "The first three days of any season rule the weather of that season." "If it rains on Sunday before mass it will rain all the week." "In leap year the weather always changes on Friday." But most proverbs are reliable short-term weather predictors for the simple reason that animals and plants are sensitive barometers to immediate changes in air pressure, humidity, wind direction, and cloud cover. Weather folklore is the poetry of meteorology: unscientific in expression but grounded in sound instinct, long experience, and reverent appreciation.

With the rise of weather science in the mid-nineteenth century, weather prognostics came under close scrutiny. Fledgling meteorologists shone the cold light of science on the old folk sayings, bent on assessing their accuracy and seeing how it stacked up against their own. One English study, which circulated to America, found that a solar halo accurately predicted rainfall on the first or second day 80 percent of the time ("When the sun is in his house, it will rain soon" is the Zuni version of this proverb). Collections of "weather wisdom" began to appear in both England and America at this time, fixing the ancient oral tradition in book form. Perhaps the most remarkable of these volumes is H. H. C. Dunwoody's *Weather Proverbs* of 1883, which was prepared under the aegis of the U.S. Army Signal Corps, one of the early incarnations of what we now call the National Weather Service. To gather material for the book, General W. B. Hazen, the chief signal officer of the U.S. Army, circulated a letter to all official weather observers asking them to record the "popular weather proverbs and prognostics" used in their section of the country "by all classes and races of people, including Indians, negroes, and all foreigners." The response was gratifying indeed, for Dunwoody was able to fill out the volume with alphabetical listings of proverbs covering the animal kingdom from ass ("When the ass begins to bray, Be sure we shall have rain that day") to wolf ("Wolves always howl more before a storm"); bird, fish, and insect proverbs ("Katydids cry three months before frosts"); proverbs describing "storm-presaging clouds" ("A mackerel sky, Not twenty-four hours dry"), plant proverbs from ash leaves to wheat, and so on. There is a wonderful section of Zuni Indian weather lore collected by the Smithsonian ethnologist residing at the pueblo ("The rainbow has but a bad character: she ever commands the rains to cease").

In his preface to the book, Lieutenant Dunwoody ("acknowledged to be the best weather forecaster in the world," according to the New York *Tribune*)[1] attributes the accuracy of many of the sayings to "the close observation on the part of those whose interests compelled them to be on the alert, in the study of all signs which might enable them to determine approaching weather changes." Dunwoody dismisses out of hand the sayings referring to years, months, days of the week, and so on ("Rain on first dog day, it will rain for forty days after," "April and May are the keys of the year"), but he concedes that as far as short-term forecasts go, even "the ablest meteorologists of to-day, aided by the most perfect meteorological instruments and the results of years of accurate instrumental observation," are scarcely more reliable than the proverbs. "It is possible," he writes humbly, "that a more accurate observation of the condition of plants or the condition and action of animals might lead to some valuable suggestion in this important field of observation."[2] Science and folklore have rarely met and mingled so gracefully. Perhaps too gracefully for Dunwoody's superiors. For his boss, General Hazen, felt compelled to preface the volume with this little disclaimer: "It is proper to add that the weather forecasts of this office are not based upon the proverbs here given, but wholly upon observations and generalizations accepted by meteorologists." This was, after all, an official publication of the War Department, as the title page declares in stern bold type. It wouldn't do, wouldn't do *at all* for readers to come away with the impression that their nation's weather was being prognosticated from some old-timer's front porch. Polish up your barometer, Private, and get that dog out of the weather station!

Turning the pages of Dunwoody's charming little book today, one feels oneself transported back to a young, green, preindustrial America—a world of water snakes and butterflies, chickens and cranes, cockleburs and jonquils. In a sense, *Weather Proverbs* commemorates the passing of this world, for by 1883, when the U.S. Government Printing Office issued the book, America was rapidly ceasing to be a nation of small farmers and was becoming more and more a society of cities and workers. Factory and office personnel generally do not get their weather from the earth and sky, but from the government via the mass media. If the National Weather Service circulated a request today for "popular weather proverbs and prognostics" used "by all classes and races of people," they would probably come up with things like: Why is the weatherman always wrong? It always rains on my birthday. The weather sure isn't what it used to be. Weather wisdom has ceased to have much relevance to our daily lives.

And yet, oddly enough, the tradition of weather prognostics remains very much alive today, with a spate of recent books collecting the old sayings and explaining how and why they "work" just as Dunwoody did over a century ago. We can choose from *Weather Proverbs: How 600 Proverbs, Sayings and Poems Accurately Explain Our Weather* by George D. Freier, *Weather Forecasting the Country Way* by Robin Page, *Weather Wisdom* by Albert Lee—to name but a few. Robin Page in *Weather Forecasting the Country Way* tells us that he relies "almost entirely on the old handed-down wisdom of country weather lore" and pretty much ignores the television babble about occluded fronts and high-pressure systems.[3] And George D. Freier, the author of *Weather Proverbs,* who has a doctorate in nuclear physics, believes that "experts" with access to high-tech forecasting equipment are "often . . . telling us exactly what the behavior of a closely observed flower or bee can tell us." A weather proverb, in his view, is "a lot of light on a subject integrated into one bright flash."[4] Weather proverbs place the weather within our grasp: they give us an intimate connection with cosmic, chaotic forces and connect us, if only faintly and fleetingly, to our ancestors. How much more satisfying it is to know that a storm is approaching because the swallows are flying low or spiders crawling on the walls ("this prognostic seldom fails," according to Dunwoody) than to get it on the Weather Channel. This chapter of our forecasting history is not over yet.

PROGNOSTICS, AS EVEN THEIR MOST FERVENT ADMIRERS MUST freely admit, have the essential limitation of being strictly local and short-term. A spider crawling up the wall in St. Louis is not going to tell me very much about tomorrow's weather here in Seattle. Forecasting on the national level for a country as large as the United States requires not only more rigorous science but also a uniform system of observation and communication. This may be obvious to us today, but it took real vision and leadership to set the first such system up and make it work. The individual who possessed this vision was an American physicist named Joseph Henry. It's strange how little known Joseph Henry is today, for in his day he was one of the titans of American science—a man who, like Thomas Jefferson, Ben Franklin, and John Wesley Powell, rose to the immense challenges that the young country furnished. In the realm of pure science, Henry's most important work was the discovery of electromagnetic self-induction, a breakthrough that set the groundwork for the invention of the dynamo and the motor. Every time you flick on one of your electrically powered machines, from the refrigerator to the vacuum cleaner, you have Joseph Henry to thank. But it was to applied science—specifically the work of scientific organization and administration—that Henry devoted the prime of his career. As the first secretary and director of the newly established Smithsonian Institution, Henry took a fairly meager budget and an ambitious but vague mission—to foster "the increase and the diffusion of knowledge among men"—and quickly gave shape to America's preeminent center for scientific research and education. Under Henry's leadership, which stretched from December 1846 until his death in 1878, the Smithsonian amassed an impressive library (later merged with the Library of Congress), laid the foundation for a national museum, sponsored journeys of exploration and surveying in the remote regions of the American West, and launched ongoing projects in archaeology, astronomy, scientific bibliography, and meteorology.

Meteorology was not only Henry's first major project as the Smithsonian's secretary, it was also perhaps his most satisfying, for in no pursuit was his success as concrete, public, and practical. In the history of forecasting in America, Joseph Henry and the Smithsonian stand at the great divide: before Henry there were proverbs, almanacs, local observations, and prognostics; after Henry, the rudiments of a national weather service were in place. His vision and determination ushered in a new age. Of course, the timing was right too.

"The present time appears to be peculiarly auspicious for commencing an enterprise of . . . an extensive system of meteorological observations, particularly with reference to the phenomena of American storms," Henry wrote in the first *Smithsonian Report* for the year 1847. "Of late years in our country more additions have been made to meteorology than to any other branch of physical science . . ." Henry seized on these "additions" and then added one bold, overarching addition of his own: the use of the telegraph—the most advanced communications technology of his day—for reporting weather conditions and making forecasts. Henry was probably not the first person to hit on the idea of weather telegraphy (some historians give credit to William Redfield, an engineer with a meteorological bent); but he was the one who made it work. Henry had himself played a key role in the development of the new technology, for he built the first practical electromagnetic telegraph in 1831. On May 24, 1844, just two years before the establishment of the Smithsonian, Samuel F. B. Morse, who perfected the telegraphic instruments, transmitted the first telegraphic message—the famous exclamation "What hath God wrought?"— on an experimental line linking Baltimore and Washington, D.C. After this the wires spread rapidly through the nation. Henry saw his chance and seized it. Soon after his appointment to the Smithsonian, he worked out a deal with the country's fledgling telegraph companies: he agreed to supply weather instruments to local telegraph stations in exchange for free use of the lines. Telegraph operators were instructed to clear their lines each morning by transmitting weather observations: the weather went out even before any death messages were sent. By 1860, Henry's system of telegraphic meteorology—the first such system in the world—covered most of the eastern half of the nation, with stations stretching from Vermont to Louisiana and as far west as central Iowa.

The other key feature of the Smithsonian meteorological project was a network of volunteer weather observers. In November 1848, Henry enlisted members of Congress to circulate notices describing the undertaking to their constituents, and the response was abundant and enthusiastic— evidently there has never been a shortage of weather nuts in this country. The volunteers were for the most part well-educated professionals—teachers, ministers, doctors, amateur scientists, with a sprinkling of farmers and artisans. Their ranks included college presidents, superintendents of city school districts, state representatives, and the president of the American Medical Association. Some volunteers had already been keeping weather journals on their own and welcomed Henry's project as a way of putting their hobby to official service. Others were eager to further the cause of science. The mid-nineteenth century was an age when science was flexing its muscles and wide avenues of progress seemed to be opening up in every direction. The Smithsonian meteorological project accorded well with the spirit of the times. Good, hard, reliable data and lots of it: there was no better way to scrub the stain of old proverbs and musty superstitions out of weather science.

Participants used calibrated instruments developed and distributed by the Smithsonian to make four observations each day (at sunrise; 9 A.M.,

3 P.M., and 9 P.M., although the hours got shifted around several times during the course of the project); they were instructed to note down temperature, barometric pressure, cloud cover, wind speed and direction, precipitation, and relative humidity, and to keep their records on standardized forms. At the end of each month, the volunteers mailed a copy of their journal to Washington, where the data was processed and analyzed by Henry's meteorological staff. There were, of course, some glitches. The demand for instruments was so great that many observers had to make do with only thermometer and barometer, while others had no instruments at all and just eyeballed it. Many observers ignored the precise instructions for filling out the forms and devised their own private systems of notation that baffled the meteorologists back in Washington. One member of Henry's staff wrote of his despair of decoding the "numerous abbreviations, new-coined characters & hieroglyphics" scribbled densely on the forms. Still, for the most part, the project rallied tremendous energy and enthusiasm for what one of Henry's colleagues dubbed "a grand meteorological crusade." At the end of 1849, the program's first year of operation, 150 people were involved, almost all of them located east of the Mississippi; at its peak, the network numbered more than 600 observers as far afield as Utah, Texas, and California, as well as in Mexico, Canada, Latin America, and the Caribbean. Henry was eager to draw on as extensive a network as possible, for, as he wrote, "in regard to climate, no part of the world is isolated: that of the smallest island in the Pacific, is governed by the general currents of the air and the waters of the ocean." To arrive at the "causes" of America's weather, the meteorologist must "study the conditions . . . of all others."

⟐   In the century and a half since Henry's "crusade," we have added satellites, computers, electronic telecommunication, radar, weather balloons, reconnaissance flights, and a far more sophisticated understanding of the laws governing the atmosphere. But our national weather forecasting system still adheres to the same fundamental structure and basic principles that Joseph Henry devised and set in motion: a continental network of weather observation stations equipped with uniform instruments at which meteorological readings are taken at prescribed intervals; the communication of weather data from the outlying stations to a central bureau in Washington; the compilation and analysis of the data in Washington; the daily dissemination of forecasts; and the issuance of storm warnings. The first national forecasting "product"—the forerunner of the hundreds of charts, graphics, computer-generated models, and coded messages that the NWS now churns out daily—was a large map of the United States and territories that Henry mounted in the great hall of the Smithsonian starting in 1856. Here he displayed in symbolic form the latest weather conditions as dispatched by telegraph from around the country and usually received in Washington by 10 A.M. The beauty of the map, as Henry wrote, was that it showed "at one view the meteorological condition of the atmosphere over the whole country." Science historian James Rodger Fleming describes it in his book *Meteorology in America*: "A piece of iron wire was driven into the map at each point of observation, and colored circular cards of about an inch in diameter were hung on the wires to indicate the weather and wind direction. White represented fair weather, blue indicated snow, black meant rain, and brown, cloudiness. Each card had an arrow painted on it and eight holes punched around the circumference so that the card could be hung to indicate the direction of the prevailing winds."[5] For a while, Henry also displayed weather signs on the tower of the Smithsonian castle, so Washingtonians in the vicinity could glance up and get the latest weather summary.

Another piece of the national forecasting process fell into place on May 7, 1857, when an enterprising editor at the Washington *Evening Star* picked up Henry's telegraphic weather readings and issued America's first newspaper weather forecast: "Yesterday there was a severe storm south of Macon, Georgia; but from the fact that it is still clear this morning at that place and at Wheeling, it is *probable* that the storm was of a local character." Now the circuit was complete: from the weather happening "out there" to local observations to Washington to news media to "weather consumer." Our country has been in the weather business ever since.

"The meteorologist is impotent if alone," wrote John Ruskin. It was largely due to Joseph Henry that America's pioneering meteorologists ceased to be alone—that they had access to each other through an ever-widening network of observers and that they were able to communicate their forecasts to an ever-larger segment of the public through the media. Three-quarters of a century before the founding of the Smithsonian, Thomas Jefferson was convinced that he could "ascertain the climate of the United States and its probable causes" if one person in each county made simultaneous weather observations twice each day and sent him their data. In a sense Henry realized Jefferson's dream, although on a slightly smaller scale. Henry also shared in the essential optimism of Jefferson's meteorological vision: both men believed that with enough precise, reliable information about the atmosphere, eventually they would "figure it out"— "ascertain the climate" in Jefferson's phrase, or as Henry put it, "solv[e] the problem of American Storms," especially the ferocious blizzards, hurricanes, tornadoes, and winter storms peculiar to North America. Weather, the ultimate expression of chaos, chained and tamed through reason and science. The Smithsonian meteorological project—and the National Weather Service that evolved from it after many changes of shape and name—was founded on this hope.

Hope was tremendously inflated in Henry's day by the hot air of controversy. Extravagant claims and furious public debate attended the birth of weather science in America. "Meteorology has ever been an apple of contention," Henry wrote, "as if the violent commotions of the atmosphere induced a sympathetic effect on the minds of those who have attempted to study them."[6] The most violent commotion raged over the nature of America's severe storms—what caused them, how to categorize them, and of course how to predict them. William Redfield, a transportation engineer who became obsessed with tropical storms in the 1820s, believed that wind was the answer to just about everything: he argued that gravity was the sole generator of wind, and that wind in turn caused changes in barometric pressure, temperature, and precipitation. The wind theory proved to be a dead end, but Redfield does get credit as the first to describe tropical storms as immense vortices of air rotating about a central "eye" of low pressure. His decade of tracking hurricanes furnished early meteorologists with invaluable observational data.

James Pollard Espy—the Storm King, as his contemporaries dubbed this colorful and outspoken man—vociferously disagreed with Redfield. Fairly early in his career, Espy latched onto the concept of convection—the notion that clouds are formed when warm air masses rise and expand, which causes them to cool and condense their water vapor—and he devoted the rest of his life to describing, elaborating, promoting, defending, and trying to prove its supreme significance, most exhaustively in his massive *The Philosophy of Storms,* published in 1841. In his view "all the phenomena of rains, hails, snows and water spouts, change of winds and depressions of the barometer" are attributable to a single law of nature: *"there is an expansion of the air containing transparent vapour when that vapour is condensed into water"* (italics in original).[7] In the Espy model, as earth-warmed air rises, winds flow in from all directions to fill the void and their inflow naturally assumes a whirling centripetal pattern. The more violent the uprush of air, the more forceful the whirlwind.

Espy and Redfield exchanged barrage after barrage of fire over their conflicting theories during the 1830s, and both in turn were savaged by advocates of other meteorological perspectives. Espy was certainly equal to battling on two or more fronts. A true product of his clamorous age, he hit the lecture circuit in 1837, preaching the new gospel of "vertical convection" to large crowds all over the country and enlisting volunteer weather observers wherever he went to provide local data. The Storm King contended that he not only held the key to the causes of weather but could actually control it—at least to the extent of making rain fall where and when he wanted. It was really quite simple. All you had to do was light a few forest fires in the West—"forty acres for every twenty miles," to be precise, "prepared and fired simultaneously every seven days in the summer . . . in a line of six or seven hundred miles long from north to south." With this weekly inferno raging up and down the West, Espy believed it was "highly probable" that "a rain of great length, north and south, will commence on or near the line of fires; that the rain will travel towards the east side-foremost; that it will not break up until it reaches far into the Atlantic Ocean . . . that it will rain enough and not too much at any one place."[8] Happily, Congress balked at appropriating the funds needed to test this hypothesis—not so much from devotion to the forestlands of the

West but more out of simple parsimony. But Espy did manage to get himself appointed to the post of "meteorologist to the U.S. government" from 1842 to 1847 with an annual salary of $2,000 paid out by the army surgeon general.

The zeal, indeed obsessiveness, with which Espy promoted his theory won him considerable fame, and not a little ridicule in his day. John Quincy Adams noted contemptuously in his diary for January 6, 1842, that he had met with "Mr. Espy, the storm breeder. . . . The man is methodically monomaniac, and the dimensions of his organ of self-esteem have been swollen to the size of a goitre by a report from a committee of the National Institute of France, endorsing all his crack-brained discoveries in meteorology . . ." This is too harsh by half. Espy's grandiloquent self-promotion and his extravagant claims, especially in regard to rainmaking, may have had a touch of charlatanism, but he was also a serious scientist (he taught mathematics at the Franklin Institute in Philadelphia) and his philosophy of storms was an important contribution to early meteorology. Subsequent advances in the field revealed that Espy's model was actually a fairly accurate description of tropical storms, such as hurricanes, but that for Temperate Zone storms he failed to take into account the importance of clashes between opposed air masses. His major flaw as a theorist, as a colleague pointed out after his death, was "his deep conviction of the truth of his theory. . . . He was not prone to examine and re-examine premises and conclusions, but considered what had once been passed upon by his judgment as finally settled."[9]

A measure of Espy's standing as a scientist was the respect Joseph Henry accorded him. When Henry organized the Smithsonian meteorological project, he enlisted Espy's help, structured the network of observers along lines suggested by Espy, and lobbied hard to get Espy back on the government payroll (the War Department had snipped his salary out of its budget in 1847). "I consider [Espy] a man of excellent character," wrote Henry, "who has done much for the science of meteorology and who in a country of so much wealth as ours should not be thus deprived of the pittance to which he was before thought entitled."[10] Henry also enlisted the help of other leading scientists of the day, including the young mathematics professor Elias Loomis, who at the age of twenty-five conducted a pioneering in-depth study of a single winter storm in December 1836, analyzing data from 102 locations in the Ohio Valley and the Northeast and drawing an early synoptic weather map charting the progress of the storm and the wind direction. "We are justified . . . in inferring that storms are subject to laws," Loomis wrote optimistically in the *Smithsonian Report* for 1848, "that these laws are uniform in their operation, and that they may be discovered . . . When we have fully learned the laws of storms, we shall be able to *predict* them."[11] Loomis felt that a maximum of three years of data collection would suffice to wrap up the "laws of storms" and solve the problem of prediction. A century and a half later, we're still at it.

JOSEPH HENRY KNEW FROM THE BEGINNING THAT THE SMITH-sonian meteorological project was too limited in scope and funding to tackle the full job of forecasting and issuing storm warnings for a country the size of the United States. Over the years he repeatedly urged Congress to step in and reorganize the project under one "comprehensive plan": a single government department with enough money (he proposed $50,000 a year) to take charge of meteorological observation and analysis for the nation. Finally, in 1870, Congress took the necessary steps—but gingerly. No one knew quite what to do with a "meteorological department" and so it was appended to the Signal Office of the War Department. There was some precedent for this association of meteorology and the military, for as early as 1814, James Tilton, the army's surgeon general, had ordered all army hospital surgeons to "keep a diary of the weather." Tilton was more interested in learning about the correlation of climate and disease than in forecasting, but the undertaking did set up a rudimentary national appara-tus for weather observations and, as expanded by his successor Joseph Lovell, provided valuable historical records for much of the eastern United States. The Signal Office was at least as logical a place to put a weather service as the surgeon general's office, for quick, efficient communication was critical to forecasting—both in the initial stages of gathering data and in the final stages of issuing outlooks and warnings. When the U.S. Army Signal Service, as the bureau was called, began doing weather in November 1870, it was a skeleton operation with a scant 25 weather stations and a meager appropriation for the coming year of $15,000—but there was one invaluable resource at its disposal: free access to the military telegraph network, which had been greatly expanded during the Civil War. The first national weather service communication ran across the wires at 7:35 A.M. on November 1, 1870. The Smithsonian meteorological project was now largely superfluous, and Henry disbanded it in 1872; the following year he authorized the transfer of the volunteer weather observers over to the Signal Service.

In many ways the Signal Service picked up where the Smithsonian left off, but there was one important difference between the two from the start: where the Smithsonian sought to solve the "problem" of American storms by ascertaining precise meteorological laws, the Signal Service focused on the more practical business of protecting life and property through timely storm warnings and reliable daily forecasts—or "probabilities" as they were called until the slightly more authoritative-sounding "indications" was adopted in 1876; the term "forecast" did not come into use until 1889. Cleveland Abbe, who had organized a local system of daily weather reports and warnings as the director of the Cincinnati Astronomical Observatory, was enlisted to serve as the Signal Service's chief forecaster and scientist. The term "chief" is a bit misleading, for "Old Probs," as Abbe was known though he was only in his early thirties, not only ran the show—he *was* the show, making all the forecasts for the entire country himself for the first few years. As a colleague said of Abbe in a memorial address, "At that time he was the only man in the country having experience in or knowledge of weather forecasting for the use of the public based upon the principles of scientific meteorology, and for some time the duty of tri-daily interpreting the meteorological observations made in all parts of the country devolved upon him alone."[12] Initially, the agency directed its attention primarily to storms blowing up along America's seacoasts and the Great Lakes, but eventually its work encompassed the nation as a whole. Specialized reports were added over the years—bulletins for farmers starting in 1873, reports on river conditions and flood warnings that same year, and frost warnings for Louisiana sugar growers in 1880, with subsequent reports targeted at growers of cotton, rice, tobacco, corn, and wheat. By 1883, the Signal Service budget was close to $1 million, considerably exceeding the sum that Joseph Henry had recommended.

Over the years, the Signal Service proved to be something of a bureaucratic rolling stone: it was upgraded to the Signal Corps in 1880; in 1891, after a good deal of wrangling between civilian weather personnel and their military superiors, its ties with the army were severed and it was handed over to the Department of Agriculture as the U.S. Weather Bureau; then in 1940 it was passed on to the Department of Commerce, where it has remained ever since, under the jurisdiction of the Environmental Science Services Administration (ESSA) from 1965 to 1970 and since then under the National Oceanic and Atmospheric Administration (NOAA).[13] The agency was rechristened the National Weather Service in 1973, and that is the name it still goes by. A surprising number of scandals and congressional investigations have erupted in its 125-odd year history, with periodic charges of corruption, favoritism, misuse of funds, and just plain ineptitude. Whenever the bureau failed to predict deadly weather with complete accuracy or with sufficient lead time, Congress and the American public demanded to know why. Some of the more outstanding Weather Service fumbles have included the hurricane that killed an estimated 6,000 people in Galveston on September 8, 1900 (the bureau warned of the approaching storm but failed to take into account the effect of the high tide); the hurricane of September 21, 1938, that ripped through Long Island and southern New England, killing an estimated 600 people (the bureau issued gale warnings four hours in advance for Long Island, but the speed and ferocity of the storm took forecasters by surprise); floods along the Ohio and Mississippi rivers in 1937 that killed 137 people (the bureau's river flood service was using outmoded equipment and forecasting techniques). Tension, frustration, and conflict are built into the agency's work. Several bureau chiefs have left or been forced from office under black clouds of suspicion and scandal. Calls for the abolition of the agency have cropped up again and again. "Weather Bureau a humbug and a grafting outfit," a handbill circulated in Michigan in 1919 declaimed. "If you are honest and work for a living, you don't want to pay for what man don't [sic] have anything to do with. The Weather is God's business only."[14]

As the agency evolved, new technology quickened the pace of communication and extended the range of observation: Marconi's wireless telegraph and the telephone early in the twentieth century, radio in the 1920s, the use of aircraft for weather observations around the same time, regular teletype transmittal of data and forecasts between weather stations by 1930. But forecasting methods and procedures remained pretty much the same during the agency's first fifty years. Observers at outlying stations sent their readings in to Washington, and bureau meteorologists plotted charts showing changes in temperature and barometric pressure, wind speed and direction, and dew point. Maps were drawn by hand with lines connecting areas of equal air pressure (isobars), and forecasts were derived by calculating the speed and direction in which high- and low-pressure systems moved. Forecasting for the Weather Bureau was a fiercely competitive undertaking in those days: promotion hinged on success and too many errors resulted in sacking. Chiefs fostered competition as a way of shaking out the inept and exposing the ungifted. Well into the twentieth century, an aura of secret magic still hovered about forecasting, as if it were a mystical sixth sense you had to be born with, like telepathy. "Very few have the faculty to forecast," Willis L. Moore, the controversial and frequently embattled chief of the Weather Bureau from 1895 to 1913, wrote to a failed aspirant, "and your record is poor even for a beginner." Chief Moore also reported to Congress that competition among his forecasters was so intense that the Weather Bureau had the highest rate of insanity of all the government departments.[15]

Despite the maddening competition, accuracy improved but little in the first half century of the weather service, hovering around the 75 to 85 percent mark—and this was for the next day's weather. Longer-range forecasts remained hardly more than educated hunches for many years. As late as 1894, H. H. C. Dunwoody of *Weather Proverbs* fame declared that if the Weather Bureau tried to extend the forecast period from thirty-six to forty-eight hours, the last twelve hours would be mere guesses. Nonetheless, two-day forecasts were introduced in 1899, and weekly outlooks in 1910, replaced by more detailed five-day forecasts in 1940. In 1941, five-day Weather Bureau forecasts had an accuracy rate of only 48 percent for temperatures and 16 percent for precipitation.[16] Tossing darts at a dartboard would have been just as effective.

The press and the public, then as now, were quick to ridicule government "weathermen" for embarrassing forecast errors and to revile them for bringing on lousy weather. One early jingle on the subject goes like this: "To check the flood you brewed, I've heard / All efforts were in vain; / Until the Bureau in Washington stirred, / And stopped the flood with a single word, / By just *predicting* rain." Greeley's New York *Tribune* reported in 1895 that "in the beginning the Weather Bureau was regarded as a huge joke, and it required many years of patient work to persuade people of its usefulness."[17] Some remain unpersuaded to this day. When the forecast is wrong, or when the forecast is right but the weather is wrong— too wet, too cold, too hot, too whatever for our taste—we blame the "weatherman." It's curious how we Americans have personified our meteorological office as the "weatherman" right from the start, erecting a straw dummy to burn in effigy at every inclemency. When a letter gets lost, we curse the Post Office, not the mailman; when April 15 rolls around, we feel the IRS breathing down our necks, not "the tax man." But when the weather turns foul, we imagine some goofy backstage meteorological Faust herding black clouds over our picnic or deliberately telling lies to lure us into calamitous embarrassment. Maybe we just don't want the government to "take over" something as private, quirky, and intimately affecting to us as the weather. The weather used to belong to us—still does in some ways —and we resent the government for meddling with it. So we mock and revile the official weatherman and we delight in any local sage, crank, or self-proclaimed prophet whose crystal ball, gut instinct, or aching knee yields more reliable forecasts than the bureau. One especially rich example of this occurred around the turn of the century when the Denver *Republican* ran the Weather Bureau indications side by side with the weekly forecasts of one Oliver P. Wiggins, a former frontier scout and associate of Kit Carson, who, at the age of ninety, predicted the weather by consulting an old battle wound in his left leg. Whitnah reports in his history that "Wiggins' efforts surpassed the official weekly forecasts to such an extent that local Weather Bureau officials requested the newspaper to stop printing these prognostications alongside those of official nature."[18] Mark Twain, credited with the famous meteorological maxim "Everybody talks about the weather, but nobody does anything about it" (which is often attributed instead to Charles Dudley Warner), could not resist taking his own hilarious potshot at official Weather Bureau jargon in this forecast parody: "probable northeast to southwest winds, varying to the southward and westward and eastward, and points between, high and low barometer swapping around from place to place, probable areas of rain, snow, hail, and drought, succeeded or preceded by earthquakes, with thunder and lightning."[19]

Twain was not laughing when he, along with thousands of other people, got caught in one of the most egregious forecast errors of all time —the legendary blizzard of Monday, March 12, 1888, that piled 21 inches of snow on New York City, 46.7 inches on Albany, 42 inches on New Haven, Connecticut, and 32 on Worcester, Massachusetts. The Signal Corps indications called for rain on Sunday and then for Monday "colder, fresh to brisk westerly winds, fair weather." It did indeed rain Sunday, and it did indeed turn colder—but then it turned sharply colder, with temperatures falling through the 20s on Monday, winds gusting to 48 miles per hour, and the light rain changing over to heavy snow soon after midnight. Ten inches had already come down before the morning rush hour got underway, and thousands of people were stranded on their way to work. The weight of snow and ice downed recently installed electric, telephone, and telegraph wires, paralyzing all rail and streetcar transport and cutting off communication up and down the entire eastern corridor. New York, which prided itself on its up-to-the-minute technology, became a dangerous frozen wasteland, its buildings dark and silent, its streets impassable to traffic and strewn with downed live electrical wires. "Great piles of [snow] rose up like gigantic arctic graves in all direction," the awestruck New York *Times* reported. "For the first time in their lives [New Yorkers] knew what a western blizzard was." "At the Mercy of the Storm, New York

Overwhelmed by a Blizzard," roared the headlines in the *Tribune,* whose lead story opened with this vivid patch of journalistic purple: "The forcible if not elegant vocabulary of pugilism supplies the phrases which will, perhaps, best reveal to the popular imagination the effect of the storm that visited New-York yesterday. New-York was simply 'knocked out,' 'paralyzed,' and reduced to a condition of suspended animation. . . . the city was left to run itself; chaos reigned, and the proud boastful metropolis was reduced to the condition of a primitive settlement."[20]

An estimated 400 people died in the storm, 200 of them in New York City. The Signal Corps's *Monthly Weather Review* for March 1888, acknowledging that the storm was "generally . . . the severest ever experienced" in New York, with drifts reaching as high as twenty feet, noted that meteorologists first detected it as a "feeble cyclonic disturbance" centered in northern Georgia. But there is no official mention of why the Signal Corps failed to predict the storm's transformation into a "western blizzard." The reporter for the New York *Tribune* paints a rather comic portrait of Sergeant Dunn of the New York branch of the Corps briefing the press on the blizzard's progress from his office at the top of the Equitable Building: "It began its 'monkey shines' in the Lake region," Dunn explains patiently for the twentieth time, "slipped away from there to the Atlantic coast; off Cape Hatteras ran into another storm that had recently started off a career of mischief from Georgia; struck with increased fury and sped away up the Atlantic Coast, howling with magnificent glee."[21] As we'll see, weather history was to repeat itself eerily 105 years later, almost to the day, when another ferocious blizzard whipped up the East Coast from the Deep South. This time, however, the NWS was considerably better prepared to deal with the storm's "monkey shines."

In many important ways, the government weather bureau made little progress between the blizzard of 1888 and the black blizzards of the dust bowl of the 1930s. The meteorologist of 1934 could communicate by telephone, telegraph, radio, and teletype (which had been introduced for use in the aviation services in 1928); he had access to data provided by over 200 weather stations as well as observations made each morning by planes flying out of Chicago, Cleveland, Dallas, and Omaha. In addition, he benefited from the strong links the United States maintained to weather services in Europe and the United Kingdom. These were solid practical advantages. But in strictly theoretical terms, meteorology in America was pretty much the same as it had been since the days of Espy and Loomis. The important pioneers of early-twentieth-century meteorological theory were not Americans but Norwegians—Vilhelm Bjerknes and his son Jakob, who started the Bergen Geophysical Institute to improve forecasting in Norway during World War I. The Bjerkneses came up with a new kind of air-mass analysis in which they likened the atmosphere of the Temperate Zone to a ceaseless battle between bodies of warm and cold air. Like invisible armies, the air masses advanced and retreated, clashing at their points of contact—weather fronts—and sending off swirling cyclones (low-pressure systems) and anticyclones (high-pressure systems). The choice of military terminology was not accidental: World War I had just ended when Vilhelm Bjerknes formalized his theory, and the image of the entrenched, blood-soaked front had been seared on the public imagination. Critic Andrew Ross in his recent book *Strange Weather* called Bjerknes's choice of the word "front" a "metaphoric extension" of World War I. The idea that storms resulted from the collision of warm and cold air masses was new (previously, meteorologists believed that low pressure caused storms); also new was the importance Bjerknes and his colleagues at the Bergen school placed on upper-air analysis: earlier meteorologists had relied almost exclusively on surface readings, but Bjerknes added a vast third dimension to the science—as high as 18,000 feet, at least in theory. "In the final analysis, the new technique of analyzing the upper air did not replace but rather supplemented the previous methods of forecasting based on surface conditions," writes one weather historian. "Surface data remained of much importance to the meteorologist."[22]

The U.S. Weather Bureau was slow to incorporate Bjerknes's theoretical breakthrough, and even when it did, the agency lacked the technology to do much with it. Upper-air wind observations by weather balloons (unmanned pilot balloons as they are known) began around the time of World War I, and during the mid-1930s radiosonde balloons came into use—balloons that radioed back continuous readings of humidity, air pressure, and temperature as they ascended up to 50,000 feet. The need for reliable detailed forecasts during World War II, especially for military aviation, pushed forecasting another step forward. By 1950, weather science had progressed to the upper limits of the available technology and theory: the fact that forecasts were better than they had been fifty years earlier was largely due to experience and more refined intuitions about how the atmosphere "behaves." As Ralph Hardy, Peter Wright, John Kington, and John Gribbin write in *The Weather Book,* "an element of subjectivity . . . continued to make forecasting as much an art as a science."[23]

Three critical innovations have given science the upper hand over art in the second half of the twentieth century: radar, which was perfected during World War II and used by Weather Bureau meteorologists starting in 1946; the high-speed electronic computer, first applied to numerical forecast problems in the 1940s and used routinely by the Weather Bureau since the mid-1950s; and the satellite, which has been an invaluable aid to forecasting since April 1, 1960, when TIROS I (Television Infrared Observation Satellite) began beaming back pictures of the earth's cloud cover from its circumpolar orbit about 500 miles in space (the first geostationary satellite GOES-1—orbiting about 22,300 miles above the equator at a speed matching the earth's rotation, was added on October 16, 1975). The computer, radar, and satellite: if you want to understand how meteorologists progressed from the botched forecast of the March 12, 1888, blizzard to the near-bull's-eye they scored when the Superstorm of March 13, 1993, roared up the East Coast, these three inventions hold the key. In the following chapter, I'll be zeroing in on this 1993 blizzard and the meteorologists who predicted it to illustrate how our national forecasting process works today.

6.

NATIONAL WEATHER NOW

EVERY FORECASTER WHO WORKS FOR THE NATIONAL WEATHER Service has a story about when he or she got the call. It almost always happened in childhood, usually between the ages of seven and ten. A tornado veered off just before it totaled their home. Lightning hit a tree across the street and they wanted to know why. Their dad got stranded by a hurricane. "The next day I knew I wanted to be a weatherman," one forecaster told me after recounting his tale of how a tornado whipped up and "just scared the devil out of me" when he was a ten-year-old boy in Texas. "And I never changed my mind from that time on." When you spend time around National Weather Service meteorologists, you hear a lot of versions of this story, but the conclusion is always the same: *I never changed my mind from that day on.* Lifelong commitment, insatiable curiosity, inexhaustible excitement over the weather: these are passions that motivate the people in charge of our national weather. These folks are not dull gray bureaucrats marking time in cinder-block Washington offices. They are people who have been obsessed for as long as they can remember with the weather—what causes it to change, how to predict those changes, how to communicate the predictions to as many people as quickly as possible. "We are all weather nuts," confessed one NWS meteorologist with some pride. "Even when we're off duty, we're very conscious of what's going on in the weather. We're always aware of what's going on out there."

Spend some more time around the NWS folks, really get them talking about what they do and why they do it for the government instead of for some television station or private weather company, and you learn that underneath the weather obsession is a real sense of mission. For them, the bottom line is *us*—our lives, our property, our safety. "I don't care if I get an attribution or not on the Weather Channel," says Ed Gross, the NWS chief of industrial meteorology (which means the liaison between the NWS and commercial weather operations such as Accu-Weather or the Weather Channel). "If they say 72 degrees and I say 83, I don't care. If the guy gets on the air and says, 'Here's my forecast,' when it's really the NWS forecast—I don't care. What concerns me is warnings. Making sure that the warnings go out, that they're consistent, that the public is aware and not confused or misled." Or as the government document entitled *Operations of the National Weather Service* states grandly under the heading "Objective": "To contribute to the safety, health, welfare, comfort, and convenience of the public, and to meet the needs of all segments of the national economy for general weather information."

Which is not to say that the NWS meteorologists don't get a tremendous charge out of the weather—the more intense, the better. Predicting whether tomorrow will be partly or variably cloudy; whether the ridge of high pressure over New England will drift east or east-northeast: totally routine. It's the severe weather—the blizzards, hurricanes, tornadoes, flash floods—that brings out the best in these people. There's a funny paradox at the heart of the career of a NWS meteorologist: your mission is to keep people from being killed by the very weather that you live for. It's a bit like a detective with a serial murder or a terrorist bomb plot. Disaster gives meaning to the job.

The lives of a great many NWS meteorologists were saturated with meaning on the weekend of March 13, 1993. This was the weekend when the Storm of the Century—the Superstorm of 1993 that appears in the opening pages of this book—hit the eastern corridor of the United States, dumping unprecedented quantities of snow from Alabama to Maine. Half the population of the country suffered through the storm; over two hundred people died in it or because of it. "The Superstorm of March 12–14, 1993, was among the greatest nontropical weather events to affect the Nation in modern times," notes the *Natural Disaster Survey Report* on the storm. "The Superstorm adversely impacted over 100 million citizens during its lifetime and severely crippled economic activities in the eastern one-third of the U.S." Yet for people inside the NWS, March 13, 1993, was a red-letter day, one they will long remember with pride, even exultation. For this was a weather disaster in which everything came together for the new, improved NWS—mathematics, physics, technology, communications, media relations. "It was a defining moment in our forecast process," notes Dr. Louis Uccellini, who was then serving as chief of the Meteorological Operations Division (MOD) of the National Meteorological Center (NMC) and has since been promoted to the position of director of the Office of Meteorology at the National Weather Service, "a process that has been evolving over the past forty years." Dr. Uccellini should know. He literally wrote the book on winter storms—*Snowstorms Along the Northeast Coast of the United States, 1955–1985,* co-authored with Paul Kocin. Clearly, there will have to be a new edition.

When I went down to Washington to find out how the NWS produces our nation's weather forecasts, it was the middle of May, only two months after the great blizzard, but already a glaring heat wave dulled the city. The foot-and-a-half of snow had soaked into the earth and evaporated into weepy Washington humidity, but the "defining moment" of the storm was still fresh in everybody's mind, perhaps freshest of all in the mind of Lou Uccellini. Uccellini and his colleagues at every level of the national weather pyramid were only too happy to talk about—indeed, to relive play by play—one of their greatest forecasting triumphs. Weather, of course, does not stop, and since the March '93 Superstorm there have been many other weather disasters—floods, tornadoes, hurricanes—some forecast triumphantly, others not. The next winter—1993 to 1994—whacked the East Coast with a whole series of snowstorms and icestorms, none of them as stupendous as the Superstorm, but all of them together adding up to one of the harshest winters in memory. But even so, the Superstorm remains a "defining moment" in Weather Service history and as such provides a good window through which to view the entire system.

I LANDED FIRST WITH LOU UCCELLINI ON THE FOURTH FLOOR of a nondescript building in the town of Camp Springs, Maryland, just a stone's throw off the Capital Beltway. From the outside the place looks like one of those suburban "professional complexes" where dentists and insurance salespeople set up shop. The one clue that something special happens here is an inconspicuous sign over the entrance: World Weather Building. No marble columns. No uniformed guards. No eagle clutching lightning bolts. No model of the globe. And yet, despite the absence of visible symbols, this building is as critical to our nation's weather as the Pentagon is to our defense. For this is the headquarters of the National Meteorological Center—the innermost nerve bundle of the central nervous system of the National Weather Service (since this book went to press, the NMC has been subsumed into the newly organized National Centers for Environmental Prediction; within this umbrella agency, the Environmental Modeling Center handles the functions of the old NMC). The National Meteorological Center is where national weather comes into focus. All the maps you see in newspapers, all those cold fronts and low-pressure systems your television weathercaster grins over, all those long-range outlooks they flash up on the Weather Channel: they all originate here. This is where the global networks converge; this is where the megacomputers are run; this is where the nation's weather *happens*: the NMC is Weather Central.

It was here, on the computer terminals of Uccellini's MOD, that the Storm of the Century began. Five days before the "event," before a wisp of cloud or a lick of wind, before so much as a smudge appeared on radar screens, Uccellini and his NMC staffers were already aware that something very strange and very powerful was brewing.

The room where the blizzard took shape, where all of our national weather takes shape, is not much larger than your average high school cafeteria, and just about as attractive. Here, seventy to eighty meteorologists work in shifts, around the clock, 365 days a year, taking in data from a network of satellites, radar installations, weather balloons, and reconnaissance planes and transforming it into "weather products" in the form of texts, statistics, graphics, and maps. These products "define the state of the weather for the next five days," as a Weather Forecast Branch fact sheet states blandly. A continent's worth of definition over a period of 120 hours: it all comes out of one cafeteria-sized room. During that second week of March 1993, while the 100 million or so people who live from Maine to Alabama were sniffing the air for signs of spring, the individuals who work in this room were watching a monster tap its beak against its shell.

"Nothing was there on Monday," Uccellini notes, handing me maps and charts showing the range of barometric pressures predicted by the five to seven numerical models that the NMC routinely runs twice each day. Each new chart jumped half a day closer to the storm. I could see that the millibar numbers were all over the place. With each jump, the models kept changing their "minds" about the intensity of the low and the path it was likely to take—one predicting rain, another placing the storm track well inland. But there was one exception to this vacillation: the NMC's Medium-Range Forecast (MRF) model. In run after computer run, the MRF model was consistent. Uccellini fans the documents out so I can see what he's talking about: "The MRF model kept showing that a complex interaction of features was blowing up in the Gulf of Mexico. This was our signal that a major storm was coming, that a powerful low was going to form over southeast Georgia on Saturday morning, and we chose to believe it and stick to our guns. The consistency of the MRF model locked us into the forecast."[1]

By Tuesday night, March 9, Uccellini's team of meteorologists became convinced that this was not going to be just another howling winter storm, but would be a "major precipitation event" depositing its heaviest snow over western Virginia. By Wednesday night, March 10, they were mapping out significant snowfall from Alabama to Maine. Three days before the storm hit, the NMC issued the first predictions for ten inches of snow along the coast, with two to three *feet* inland. And now the phrase "historic proportions" began to appear in NMC communiqués. "This was unprecedented," says Uccellini, getting swept up once again in the drama of the storm. "A few years back we would never have predicted more than four inches this far in advance. With this storm we had unprecedented lead time based on forecast models. We had unprecedented accord with local offices, and unprecedented effort to get the media plugged in to alert the public." By dawn on Friday, the ordinarily terse, colorless Weather Forecast Branch bulletins were screaming like tabloid headlines: "A great storm is approaching . . . monster low [moving] up the East Coast . . ." "One of the great winter storms in Eastern U.S. History . . ." Local forecast offices up and down the East Coast began issuing blizzard warnings on Friday afternoon for late Friday night and Saturday. If you had your radio on, you heard the hair-raising bleeps of the Emergency Broadcast System—"This is not a test . . . repeat . . . this is not a test"—breaking into regular programming to issue blizzard and flood warnings. Ominous blizzard predictions kicked off local evening news shows in Washington, New York, Philadelphia, and other East Coast cities. "If you're a meteorologist, you know it's going to be one of the major East Coast storms of all time," well-known Washington meteorologist Bob Ryan bubbled on WRC, an NBC affiliate. "When those NWS guys start using terms like 'historic proportions,' 'record-breaking,' and 'extremely dangerous,' you know this one will come through," John Curley promised viewers on Washington's Channel 8. All systems were go: the entire East Coast was counting down to the blizzard as if it were a rocket launch.

And yet, Uccellini admits that when people from the media swarmed into the NMC on Friday for a briefing, he was "sweating bullets" of anxiety that his office might have overhyped the storm. "CBS wanted me to call it the Storm of the Century, and in fact the phrase has been attributed to me," he says. "But I refused to do it. The century still has seven years to go." (Wise restraint, as it turns out, given the ferocious series of winter storms that hit the East Coast the following winter.)

Jim Belville, meteorologist in charge of the National Weather Service Forecast Office for Washington, D.C., located in Sterling, Virginia, was also sweating bullets that Friday. "On Friday night, I went home at 4 P.M. so I could get to sleep early, because I knew I'd be in here at 4 A.M. the next day," he told me. "At around 9 P.M. I went outside to look at the sky —it was clear and 40 degrees, and I remember thinking, 'Jeez, it's not cold enough.' One model showed rain, and I wondered if we had bit off the wrong model. I dreamed that night that the storm came and it was all rain —this is the truth—and I woke up at three in the morning in a cold sweat. Then I looked outside and there was already three inches of snow on the ground and it was still falling. And I knew we had nailed it."

Uccellini, Belville, and their colleagues in the NWS always smile when they get to this part of the story. Just two weeks earlier they had "gotten killed" when a predicted major East Coast snowstorm never materialized. The models disagreed, and they had picked the wrong one. But not this time. This time they really did nail it.

By the time this combination blizzard/hurricane headed out to the Canadian Maritimes late on Sunday, no American with access to a television or radio was in any doubt about just how hard and how straight the NWS nailed it: At least 219 deaths from Cuba to Canada (other estimates run as high as 270—more than three times as many people as hurricanes Hugo and Andrew killed together). Property damage estimated at over $2 billion. Fifty inches of snow on Mount Mitchell, North Carolina; 47 inches in Grantsville, Maryland; 43 in Syracuse, New York; 13 in Birmingham, Alabama. Record low pressure in Worcester, Massachusetts, Washington, D.C., Richmond, Virginia, Raleigh, North Carolina. Three million people without electricity. More than fifty tornadoes in Florida's panhandle and a nine-to-twelve-foot storm surge along Florida's Gulf coast. Winds of 135 miles per hour reported on New Hampshire's Mount Washington. For the first time ever, a single storm shut down every major airport on the East Coast. Altogether, meteorologists estimated, some 44 million acre-feet of water (one acre-foot is the amount of water that would stand one foot deep over one acre of land) had fallen during the storm—more water than flows out of the mouth of the Mississippi River in forty days (forty normal days, that is, not forty days of the record-breaking floods that would swell the Mother of Rivers later that same year).

"What would have happened if the storm had fizzled?" I asked Uccellini when he got through reeling off the statistics. "I admit I was a nervous wreck Friday night," he replied. "If it had turned to rain at Pittsburgh, it could have been the bust of the century. If this forecast was wrong, after all the hype, we would have gone out the window. Literally. It would have set us back ten years. After mispredicting an event like this for the East Coast, we'd need a decade to get our credibility back."

"So you were lucky," I ventured, hoping to get a rise out of him. Uccellini's answer, to boil it down drastically, was yes and no. Yes, because luck, intuition, going with your hunches—whatever name you want to give to the extrascientific—remains inextricably intertwined in the science of meteorology. No, because the consistency of the MRF model, the foundation on which this historic forecast rose, was dictated solely by the numbers and the formulae programmed into the model. Everyone I spoke to at the NMC was preoccupied in one way or another with numerical models: designing them, running them, comparing them, refining them, knowing which one to choose. This is the NMC's core function today—the function that happens on a national level here and nowhere else. As Dave Olson, chief of the NMC Weather Forecast Branch, puts it, "The stuff that drives all the television guys, the whole weather business in the U.S.—it all depends on the numerical models that come out of here. If we went out of business, they all would be out of business too."

So if I wanted to understand how the NWS nailed the Storm of the Century, I'd have to understand the basics of numerical modeling. Again, Uccellini was my enthusiastic guide. The atmosphere is "a continuous fluid," he explained to me, set in constant motion in three dimensions by the immutable laws of physics. But "continuous" is not within the capability of our science or technology. When the meteorologist sets out to measure the behavior of the atmosphere at a given point in time in order to predict how it will behave in the future, he is forced to discretize—to break up the continuous fluid into measurable chunks. And that, inevitably, introduces errors. "The whole trick is to minimize these errors," says

Uccellini. In numerical modeling of the atmosphere, you minimize errors by enhancing the accuracy of your data—the readings transmitted from weather stations as well as from satellites, reconnaissance flights, ships, weather buoys, and from the 98 weather balloons that are launched simultaneously twice each day around the world—and by refining the resolution of your grid. In other words, the smaller the cubes into which you break down the atmosphere, the better the results. When the NMC began numerical modeling about thirty-five years ago, the grid points were spaced 381 kilometers apart; the range on currently used models is between 80 and 180 kilometers, but the NMC is now experimenting with models with resolutions of between 30 and 40 kilometers. Critics note that weather models rely on only the most basic and easiest-to-measure variables—temperature, humidity, and air pressure—totally ignoring such crucial factors as sea-surface temperature, evaporation, the impact of topography on wind speeds. And even an 80-kilometer grid still falls far short of "continuous." But even so, numerical models represent a giant step forward for meteorology. As James G. Howcroft, deputy director of the NMC, told me, "The numerical models are the reason our five-day forecasts are now as accurate as our three-day forecasts used to be, and our 72-hour forecasts are as good as our 36-hour forecasts used to be." At present, the NMC runs its suite of models twice each day—timing the runs around the collection of data from the weather balloons that are launched daily at 7 A.M. and 7 P.M. Each model has its own set of formulae that manipulate the data in slightly different ways. Basically, what the models give you is a simulation of atmospheric change: you plug in the numbers (temperature, humidity, and air pressure), and the model uses the fundamental laws of physics to calculate how the numbers will change in the future. Meteorologists use model output in many forms; but for the layperson, the easiest way to visualize a model is to look at a map it has created: the map will show predicted atmospheric pressure change in twelve-hour increments over the next several days. By putting the maps together, you see how pressure systems move. The Superstorm of '93 offers an especially clear example.

The theory behind numerical weather modeling grew out of the pioneering work of Norwegian meteorologist Vilhelm Bjerknes around the turn of the century. And the basic mathematical concepts were in place by World War I, formulated by British scientist Lewis F. Richardson in his 1922 book *Weather Prediction by Numerical Processes.* The major roadblock was that there was no way to put the theory into practice: by the time you executed the staggering number of mathematical calculations needed to run even the crudest model, the weather system you were trying to predict would have been history. Numerical modeling became feasible only after the electronic computer was perfected following World War II. The first task that computer pioneer John von Neumann and his colleagues at Princeton's Institute for Advanced Study assigned the ENIAC, the original electronic computer, was a numerical forecast. The first numerical forecast, carried out by moving plugboards and working a huge set of keys by hand, was run in April 1950. It took nearly fifty minutes for the program to digest data gathered from 768 weather stations. By 1953, von Neumann's machines could carry out a twenty-hour numerical forecast in six minutes. Weather forecasting and computer technology were paired right from the start, and these two applied sciences have been closely associated ever since.

Obviously, it takes one monster of a computer to crunch through the suite of different models that the NMC currently runs twice each day. The NMC's reigning monster is the Cray YMP-832, a supercomputer capable of performing approximately two billion operations per second and housed in pristine, climate-controlled splendor a few miles from the World Weather Building in Suitland, Maryland. Each time it runs a model, Cray starts with the current readings and then, using a different set of equations for each model, calculates how these readings will affect the weather in ten minutes, and then in ten minutes after that, and so on, in ten-minute increments out to forty-eight hours. During one computer run each day, Cray runs predictions out as far as ten days into the future—or, as meteorologists say, out to day ten. According to one estimate, it would require 123,000 human brains working at maximum efficiency to carry out the calculations that Cray does at the flick of a switch. Numerical models and the ability to run them on supercomputers are probably *the* greatest advance in meteorology since the launching of the TIROS satellite in 1960.

Uccellini is quick to point out that the numerical models that Cray whips out are not *forecasts* but *guidance* for local meteorologists to use in making their area forecasts. In fact, the thousands of texts, maps, charts, and graphics that Cray produces each day are known collectively as the "national guidance package." These are in essence technical briefings on how the big weather picture is shaping up: the ridges of high pressure drifting westward over the Plains; the squall lines erupting along the boundary between hot and cold air in the Midwest; the upper-level disturbance over the Carolinas. For example, on the morning before the blizzard of 1993, one piece of the guidance package, a "Quantitative Precipitation Forecast Discussion," looked like this: "XPC EXPLOSIVE DVLOPMNT OF SFC LOW UP THE ERN SEABOARD BEGINNING VRY LT FRI NGT AND CONTG SAT/SAT NGT . . . AS LARGE AREA OF −40C AIR AT H5 ACCOMPANYING NRN STREAM S/WV DIGGING SEWD THRU THE NRN ROCKIES/NRN PLAINS THIS EVE GETS INVOLVED WITH INITIALLY RATHER FLAT SRN STREAM WV COMING OUT OF THE ERN GULF." This might not mean a great deal to the layperson; but to a meteorologist it was a crucial stage direction for the imminent drama of the storm.

When Cray runs through a suite of models, it seldom presents the meteorologist with a neat row of identical numbers. The various models frequently diverge, sometimes substantially. When this happens, the ball is back in the meteorologist's court: he or she has to compare the computer-generated numbers and decide which set is the best. The NMC guidance package is thus a "man-machine mix," as one official Forecast Branch document puts it, in which "the final products are superior to what either man or machine operating alone would produce." What man contributes to the mix is a compound of experience, intimate knowledge of how the models behave, subjective "gut feelings," and a touch of "black magic," as one NMC staffer joked. Forecasters, like generals surveying their arsenal, note that each model has its own quirky strengths and weaknesses—one consistently underestimates summer rainfall; another tends to muff up on the speed at which weather systems move; yet a third has a good track record on autumnal fronts but shows a blind spot when it comes to major winter storms. And, as befits the strange behavior of weather, the models perform erratically over time—any given model will match weather patterns closely for a spell and then deviate as the weather changes. But, as Uccellini puts it, "despite the frequently large differences between the models, forecasters more often than not select the correct one. We know when a given model is going off the reservation." "We never pick the worst model," adds Olson, "though we don't always pick the best."

NMC staffers have felt compelled to publish papers defending their "subjective" input in choosing among conflicting models—as if the human element is something they have to apologize for.[2] But some meteorologists believe that automation has already gone too far and that forecasts are beginning to suffer as a result. As one television weathercaster told me, "As supercomputers resolve the atmosphere to finer grid scales, and as we run more models and run them out longer, there is no question that medium-range forecasts have improved dramatically. Fifteen years ago we didn't have a clue about days three to five—now we're even going out to day ten. But the forecast for tonight and tomorrow has not improved at all and may actually be getting worse. There is too much emphasis today on computer models and not enough on the individual meteorologist's subjective gut sense. Some of the worst storms I've seen have not shown up on the models. And ASOS [the new automated surface observing system in which readings are taken by automated weather instruments and transmitted electronically] has wonderful aspects, but not having a pair of human eyes to see the types of clouds and the types of lightning is a real loss." The NWS is aware of these criticisms and is taking steps to address them. As one NWS staffer told me, "The new direction that we're moving in is to provide more and more detail for the 0 to 24-hour range. This shift of emphasis to the small scale is really a revolution for the meteorologist. Today's weather is dominated by a little eddy in the lee of some landform. This doesn't even show up on the model grids. But this is what we're directing our attention to in local forecasting. Our improvements should soon be obvious to the user" (meaning us).

As you might expect, given the nature of their obsession, the NMC people are always chafing to squeeze more out of their models and their computers. "We always want to push the limits," insists Uccellini. "We always want to run the models four times instead of twice a day. But for that we need bigger capacity and faster machines." He would also like to see more refined models—models that break down the atmosphere into smaller and smaller units, so as to approximate more and more closely the "continuous fluid." Bigger and faster and more refined, of course, cost money. And money has not been flowing very freely in Washington these days. But even if Uccellini could push the limits of science and technology as far as he wanted to, he acknowledges that eventually he would run up against a meteorological impasse. Most meteorologists agree that the barrier now stands at about day fourteen: when you run a model out beyond this, the deviation becomes so great as to render the forecast practically useless. Common sense and a coin toss would be as likely to predict the weather two weeks hence as a model run through Cray.

The NMC does issue thirty- and ninety-day outlooks of temperature and precipitation averages through its Climate Analysis Center, but these are developed largely from statistical comparisons based on historical trends, with numerical models used only for the first ten days or so of the period. For example, CAC meteorologist Russell Martin told me in October 1993 that the next thirty days here in Seattle were likely to be warmer and drier than average: numerical models indicate that the spell of remarkably warm dry weather we'd been enjoying for the past few weeks was likely to continue during the first week of October, and statistics generated by comparing current conditions with historical records of upper-air height patterns gave the warm dry trend a better than average chance of hanging in for the remainder of the month. As Martin explained, "We compare observations from last month with how similar patterns have persisted in past years. Then we derive temperature and precipitation outlooks from this pattern. But it's not a one-to-one correspondence—other factors enter in." Martin acknowledged that the CAC method has a serious underlying limitation: complete records of upper-air height patterns are only available back to 1948—a minuscule data set. "There is no question that our accuracy would improve if we had a greater data set," he says. "As things stand, our outlooks are only slightly above random chance." So, in the thirty-day precipitation outlook, a coin toss would give you a 33 percent success rate (remember that you have only three alternatives—normal, above normal, or below normal), whereas if you listened to the CAC you'd up your odds to about 40 percent. But even a small edge is better than none when the numbers are large enough—for example, a huge fuel oil company allocating millions of dollars in resources for the upcoming winter season. Russell Martin puts it this way: If you want to plan a picnic for two weeks hence, the CAC thirty-day outlook of below normal precipitation is not much guarantee of dry weather. But if you're a professional picnic planner, setting up thousands of picnics in your region summer after summer, then you'll come out ahead over the long run by listening to the CAC. It's still a roll of the dice, but with loaded dice.

On the cutting edge of numerical modeling, NMC forecasters are experimenting with a concept called "ensemble forecasting." In essence, this involves running a model repeatedly—anywhere from ten to forty times—with small variations in the input conditions. The meteorologist then compares the results, and uses the consensus or divergence as the basis for a forecast. "Ensemble forecasting is an attempt to quantify uncertainty," explains Dave Olson. "Let's say you're working on the probability of precipitation for this Sunday. If your model runs come up with a wide range of solutions, then the probability is low. If the solutions fall into a narrow envelope, then you would up the probability." Meteorologists are also reanalyzing old weather maps using current models to try to pinpoint exactly where the models wander off into uncertainty. These experiments may well result in improvements in the reliability and accuracy of our daily and long-range forecasts—but subtle improvements. Nothing on the experimental horizon is likely to be as dramatic as the implementation of numerical modeling a generation ago.

BACK TO THE STORM OF THE CENTURY AND HOW IT WAS FORE-cast. As we've seen, Uccellini and his folks got wind of the storm when run after run of the MRF model indicated the same basic scenario. And each day during the week before the blizzard, the NMC marshaled its findings in the maps, charts, and texts that make up the national guidance package: the Big Picture. But the mosaic of Little Pictures, the actual forecasts that specified how much snow would fall in your area, when it would start and whether it would change to rain, how hard the wind would blow and from what direction—all of these were created locally, at the dozen or so Weather Service Forecast Offices (WSFOs) situated in the storm's path. It's worth mentioning that the status of these WSFOs is currently in flux as the National Weather Service modernizes. At the moment, there are 52 WSFOs covering the country—about one for each state, though Alaska and Texas rate three offices each because of their size and meteorological diversity, while the Boston office covers all of southern New England. One rung down are the 250-odd Weather Service Offices (WSOs), where observations are made and weather information disseminated. Under the new NWS modernization program, the WSFOs and the WSOs are being merged into a national network of 119 WSFOs, all equipped with the latest technology.

Jim Belville is the meteorologist in charge of one of the modernized WSFOs, a showcase operation and a model of what the NWS will soon look like everywhere. It is, of course, no coincidence that Belville's office, located a few miles from Dulles Airport in Sterling, Virginia, was chosen for updating: this is where the weather forecasts for the nation's capital are made. The other two showcase WSFOs are located in Norman, Oklahoma (the heart of tornado country and the site of the National Severe Storms Laboratory), and in Melbourne, Florida, which serves the Kennedy Space Center and prime hurricane country—two other areas that are frequently in the meteorological limelight. But the beam of public scrutiny seems to fall most intensely on Sterling. When a President is being inaugurated, when the country turns out to celebrate Memorial Day on the Mall, when the air force wants to include fighter jet flyovers in some celebratory parade —Jim Belville is the man in the meteorological hot seat.

That seat was very hot indeed the weekend of the blizzard. "That weekend, Secretary of Housing and Urban Development Henry Cisneros was planning a conference, and I told him to cancel it," recalls Belville, a large, affable, modest man whose accent and manner are unmistakably southern. "You become *very* visible when you're at that level. The hardest thing is to back your way out of a bad forecast." By "bad" he meant inaccurate—not lousy weather. One can imagine that it might have been a touch awkward on Monday morning, if the Storm of the Century had fizzled, to explain to Secretary Cisneros that the MRF model had crunched the numbers wrong and sorry about the conference.

Belville explains that the essence of what the WSFO does is to take the NMC guidance package and tailor it to a specific region or zone. In the case of the Sterling Forecast Office, that region includes not only the Capital District, but parts of surrounding Maryland, Virginia, Delaware, and West Virginia. The team of meteorologists who work in round-the-clock shifts at Sterling have on-line access to the full range of NMC products: the daily weather maps, model runs, precipitation probabilities —all instantly available at the click of a few keys. "We take their guidance and adapt it to our local area with modifications," Belville explains as we tour the Sterling "war room," a sleek, cavernous, dimly lit chamber in which computer monitors glow like votive candles. "We use radar to validate what the models are showing. For example, for the blizzard, the NMC guidance put the rain/snow line in the western part of our area—but based on our knowledge of local conditions we moved it east. The NMC advised us to reconsider and go with a warmer model, which would have meant rain. But we decided to stick with the colder model. We have a multitude of microclimates in this region, and we can take these local conditions into account in a way that the national guidance package can't." In a sense, the business of the WSFO is to customize the NMC products. Again, there is a man-machine mix, but here on the local level man holds even greater sway. A person, over time, just gets the "hang" of how weather behaves in his or her region in a way that no machine has yet been able to match.

In the case of the March blizzard, once the Sterling forecasters became convinced that the storm really *was* going to hit their area hard, Belville began spending more and more time alerting key people in the media and the community. "The difference between us and the folks over at NMC is that we deal directly with users at the local level," he says. "We must make meteorological science accessible to the layman and to the state and local official. We also interface with the media, which is the number one conduit of information to the public. For that storm, we started dealing with state officials on *Wednesday*—four days before the blizzard actually hit. We had that much confidence in the forecast models. I called them and told them there was 'potential' for a severe winter storm. On Thursday we put out the initial storm watch."

Ordinarily, Belville's office issues four public packages a day, updating the forecast for his region every six hours; but for this blizzard, the Sterling meteorologists stayed in much closer contact with the local news media. "By Thursday, local TV and radio meteorologists were already under a lot of pressure from their news directors and from the public," he recalls, "and when we issued the watch this far in advance, it took some of the pressure off them. But given the magnitude of the potential threat to life from winds, snow, and flooding, people needed extra preparation time. From here on I personally communicated daily with the TV stations. We wanted everyone to be telling the same story."

Like Uccellini over at the NMC, Belville got tenser and tenser about all the hype as the hours before the "event" dwindled away. "We never really considered backing off our forecast, though I must admit there was still a certain amount of doubt as the weekend approached. Some private forecasting companies were starting to back off from the snow forecast. The NMC models were varying a bit and confidence levels were not as high." But despite the tremors of doubt, Belville continued to plan for major weather. "On Thursday I bought food for my family, found the fireplace grate and flashlight, and I told them, 'You're on your own.' I spent Friday talking with the media and that's when we made the decision to go out with a blizzard warning. This was the first time such a warning was ever issued in the Washington, D.C., area." Belville went to bed early on Friday in anticipation of the long haul ahead. While he sweated out a nightmare in which the storm turned out to be all rain, the night shift out at Sterling was jubilantly watching their forecast come true. "By midnight Friday, when our meteorologists began picking up wind speeds on the radar, we knew we had bitten off the right model. The meteorologists on duty were ecstatic. There was no longer any question in their minds."

Their ecstasy was somewhat tempered by the fact that they were trapped at work: the forecast office is about ten miles off the highway down a two-lane rural road, not a high priority for snowplows. Belville had taken precautions, topping off fuel tanks as early as Wednesday and arranging for a two-day supply of food to be brought in, as well as bedding. By four o'clock Saturday morning, with three inches of snow on the ground and more coming down fast, he was back at Sterling, where he and a staff of fifteen remained for the duration of the storm. "The place went bananas with phone calls," he recollects, getting swept up again in the unfolding drama. "I got a call from the Secretary of Commerce's office—Ron Brown wanted to be briefed during the storm. I was somewhat leery of calling him at home, but I did. We talked eight times during the event. I had to put together a briefing package and fax it to the President during the event —that took some time to assemble. I did 125 radio interviews in an eighteen-hour period. I was on the phone almost continuously from 6 A.M. to 9 or 10 P.M. This was incredible." Meanwhile, outside the station winds were gusting at 40 to 50 miles per hour and the snow and sleet kept piling up—16 inches at Sterling and 12 to 13 inches in downtown Washington by the time the storm ended on Sunday. "Armageddon!" howled CNN. "The mother of all blizzards." At one point during the height of the storm, one especially hyper newscaster dubbed it the "Storm of the Millennium," utterly disregarding the fact that most of the millennium remains a meteorological mystery. But nobody would deny that this was a storm to remember.

Least of all Jim Belville. He ended up spending thirty-six straight hours in his office that weekend, sleeping on the floor and eating cold food with his colleagues. When it was all over and the roads were passable by midday Sunday, he went home and crawled into bed with a raging fever and a full-blown case of the flu. But the media still wouldn't get off his back. The Storm of the Century was the news story of the moment—just as good as a juicy sex scandal or a terrorist attack. CNN called and wanted him to hustle down to their Washington studio. Fox News asked him to comment on whether the Clinton administration's handling of this national weather emergency was any different from the way the Bush administration had coped with major storms. The producer of *Face the Nation* wanted him to appear on a show about global warming. Belville refused all of them, most emphatically the last—and he would have refused even had he not been "sick as a dog." "I worry about tomorrow's weather," he says, "not climate change." And even after a weather event of the magnitude of the Storm of the Century, there is still tomorrow's weather to forecast. Meteorologists, like newspaper reporters, always have a new story to cover —a different one every day.

WHEN WE GOT THROUGH TALKING ABOUT THE BLIZZARD, Belville showed me some of the nifty new gadgets that today's meteorologists get to play with for a living. One of my favorites was the National Lightning Detection Network: on a video display terminal you can see where lightning is flashing anywhere in the country, just a heartbeat after it has actually struck. While we chatted, lightning bolts were landing in upstate New York, near the Canadian border, and we saw the storm moving southeast. This, to me, was supremely thrilling. If I had one of these in my living room, I'd be glued to it—I'd probably call people up where the screen was flashing to see how bad the storm was. According to Belville, I'd never be bored: "There is always a storm somewhere in the country, even in winter." A couple of thunderclaps sounded over the Washington region during the Storm of the Century, on top of the snow, sleet, and wind, so the National Lightning Detection Network came in handy that weekend right in Belville's own backyard.

The piece of new equipment that Belville and his staff are most excited about is their Doppler radar, officially known as WSR-88D (which stands for Weather Surveillance Radar, developed in 1988, Doppler). Doppler radar is the centerpiece of the modernization that is currently transforming forecast offices nationwide: when that process is finished, all 118 forecast offices will be equipped with Doppler. As James G. Howcroft at the NMC put it, "This is the principal technique that is going to change the way we live." Belville and his staff have had the Doppler for a while now, but they're still not taking it for granted. Once you get them started on the subject, they just can't contain their enthusiasm. What these forecasters love most about the Doppler is that it adds an extra dimension to their detection of weather systems. Whereas the conventional radar that the NWS (then the Weather Bureau) has been using since 1946 indicates the position and density of clouds, Doppler radar *also* measures the velocity and direction of movement of a weather system as a whole and of the winds inside the system. With Doppler radar, a meteorologist can tell whether rain-bearing clouds are moving toward or away from their weather station and how fast; they can determine the power of a thunderstorm and the total amount of moisture enfolded in a cloud; they can assess how much precipitation is falling at any given moment. Doppler can even pick up winds and warm/cool air boundaries that are rippling through cloudless air. Doppler radar's ability to detect wind direction and velocities makes it invaluable in pinpointing tornadoes and other violent, fast-moving storms. "Doppler has made a tremendous difference in nowcasting—how you track the formation of storms in the short term," says Gary Shore, a television weathercaster at KJRH in Tulsa, Oklahoma, which has had its own Doppler radar set since 1984. "But you have to understand that Doppler is very complex and the Doppler imagery lends itself to many different interpretations."

Doppler radar, as the name indicates, takes into account the Doppler effect—the apparent change in the frequency of a wave that results from a change in the distance between the source of the wave and the receiver. The classic example of the Doppler effect is the train whistle that sounds like it has a higher pitch when the train is approaching and a lower pitch when it's receding. Doppler radar measures the change in the frequency of a cloud's water droplets as they move toward and away from the radar position; conventional radar shows the *intensity* of precipitation but not the direction in which the storm winds are blowing.

"One of the real beauties of the Doppler is that it gives us the internal structure of a weather system," explains Belville. He and his meteorologists have used Doppler to detect potentially destructive weather fronts speeding westward toward the Chesapeake or ripping off the Atlantic. With record-breaking outbreaks of tornadoes in Maryland and Virginia during 1994, the Doppler has not been sitting idle. On November 1, 1994, the Doppler enabled Sterling forecasters to see a tornado spinning off a line of severe thunderstorms ripping through Baltimore. "We got the tornado warning out twenty to twenty-five minutes before it touched the ground," says Belville. This simply would not have been possible with conventional radar. "The Doppler is still exceeding our expectations," he concludes.

Since it was a slow morning out at Sterling when I visited (only 40 percent chance of thunderstorms late in the afternoon), I got to spend as much time as I wanted watching Jim Weismueller, one of the staff meteorologists, demonstrate the Doppler's bells and whistles. Weismueller loads in an optical disc that had stored data from a whopper of a storm that hit the area on May 5, 1993. He pokes a few keys, and presto, clumps of storm cells, color-coded for intensity, blossom on the screen. Numbers flash up the wind speeds, and a blinking light indicates the likelihood of hail (there is also a wailing alarm that goes off at the approach of severe weather). Click: the county and city names appear. Click: the network of highways spider out over the map. Click: we zoom in on the most destructive storm cell. Click again: the image rotates to present a vertical cross section of the storm, which is stratified with air layers of different temperatures and wind speeds. "The structural cross section shows us the potential for high wind gusts or large hail," Weismueller explains. "See that"—he points at a slice of the most active portion of the storm cell. "That's a potentially dangerous area—there are strong updrafts with a strong inflow of air. We assumed from that that we'd get large hail, so we issued a severe thunderstorm warning."

"It gets really interesting in bad weather situations," adds Belville, making me wish that a supercell would sail in that very moment. (Strangely, my wish came true the following afternoon when a vigorous squall line whipped off the Chesapeake with gale-force winds and driving rains. It toppled trees and killed one woman as she was driving home from work. I was caught in the storm while trying to get from a metro stop to my hotel—a five-minute walk—but even with an umbrella I had to take cover until the worst of the rain blew through. As I stood watching the torrential downpour, I kept thinking how the Doppler radar screen out at Sterling must be flashing like a Christmas tree.)

Eventually, all of America's Weather Service Forecast Offices will be equipped with Doppler radar and the other new technologies (wind profilers, automated instruments for measuring temperature, humidity, and air pressure, a new data-processing system) that are part of the NEXRAD (Next Generation Weather Radar) modernization program. The NWS goal is to have Doppler cover the entire country below 10,000 feet (with a few shadows in certain remote areas of the Rockies). But it remains unclear when that goal will be reached. Everyone admits that the modernization is long overdue: the 1950s-era weather radar has been in place so long that the NWS can't even get parts for it when it breaks down. The NEXRAD program, like most federal programs, has had its hiccups during the implementation stage. Even before Doppler radar is up and running nationally, critics have moved in to attack the new technology, pointing out that it probably won't work as well in mountainous areas as it does on the relatively flat terrain around Washington, D.C., Oklahoma, and Florida and that, like all radar, it is inept at imaging snowstorms. There were also allegations of political impropriety in the choice of UNISYS to build and install the new technology. When questioned about Doppler's limitations, the guys out at Sterling just shake their heads and smile. They've seen what it can do in action. They wouldn't go back to the old radar for the world. But of course the ultimate judgment rests with us: our lives, our safety, our property, our comfort. We're standing by.

IT'S CLEAR THAT YOU HAVE TO LOVE WHAT YOU DO—AND TO love weather—if you're going to work for the NWS. But much as you love it, you also have to learn to laugh about it, because once people find out you're a weatherman, they can't resist taking potshots at you. "Maybe you guys would do a better job if you looked out the window for a change."

"Why do you always have to give us bad weather on weekends?" "Can't you *do* something about this weather?" "Chance of rain thirty percent—what is this, a horse race? Anyone taking bets?" The person in the street doesn't really care about the $3 billion NWS modernization program. All we're really interested in is tomorrow's forecast—and when the weatherman blows that, we throw up our hands and laugh, just as we used to laugh about the Weather Bureau and before that the Signal Service. "It's endless, the weatherman jokes," says Andy Stern, a meteorologist in Belville's office. "You just have to have a sense of humor. You get used to it. People always notice when you're wrong and they never forget it, especially the media. But nobody ever remembers when I'm right."

Some weathermen are better at rolling with the punches than others. Lou Uccellini took time out from our interview to show me a videotape he had made of a CBS news broadcast the night before the March blizzard. While the standard TV news footage of wind-driven snow and skidding cars flickers across the screen, the voice of reporter Jacqueline Adams reels off the terrifying blizzard warnings that the NWS has issued for the entire East Coast. Then Adams herself appears on-screen: "This could be the biggest and meanest storm ever," she tells the camera, emphasizing the *could.* Significant pause. "Of course, the NWS has a long history of being wrong." Uccellini writhes at this part, although he has clearly seen it a dozen times—in fact, he has taped it for the express purpose of making himself writhe. "This is what we're up against," he storms. "It gets you right between the eyes." The serious side to the jokes and digs, Uccellini says, is that they undercut the warnings that the NWS and local emergency managers put out together. Predicting a major storm four days ahead of time, getting the word out to the community, issuing storm watches and warnings—it's all a monumental and expensive waste if people are going to thumb their noses at you. Don Wernly, chief of the NWS Warning and Forecast Branch, notes that at its outer edges, the work of the NWS shades into public relations and psychology. Part of the NWS credibility gap is a failure to communicate its successes. How many people are aware that in just a decade the NWS has upped its range of accuracy by a full day, so that their five-day forecasts are now as accurate as their four-day forecasts were in the 1980s? Or that the accuracy rating on forecasting atmospheric pressure out to thirty-six hours is 95 percent today compared with 35 percent in 1955? Does anyone even remember that before 1940 there was no such thing as a five-day forecast? Our skepticism and suspicion run deep. Even though we pay for the NWS with our taxes—and get a "damn good bargain" for it too, according to NMC deputy director Howcroft, who points out that each taxpayer shells out only about two dollars a year on the 4,200-odd NWS employees—deep down in our bones we still think we know better. The NWS has an image problem, there's no doubt about that.

If it's any consolation to the NWS, I personally had a good deal more respect for the operation and the people inside it after my trip down to Washington. It was their weather mania that appealed to me most. Back home, mulling over my interviews with these professional government weather freaks, I was reminded of the fact that the Weather Service has its roots in the military and in medicine—in fact, army surgeons were our nation's first official weather observers. Some faint glow of that military-medical aura still pervades the NWS today. The obsession with high-tech gadgets to gather increasingly minute and accurate data. The long stretches of dull routine punctuated by explosive crises. The dominance of practice over theory, at least on the front lines. Maybe that's it most of all —the fact that meteorology, like medicine and warfare, is fundamentally an applied science, that it lives or dies on the success of its application. Like the best doctors and the best soldiers, the best weathermen are masters of machinery but they also know when to go with their gut. As Jim Belville puts it, "We've got very good technologies and they keep getting better, but the bottom line is that the meteorologist still makes the decision." And in making the decision, the weatherman still wakes up in a sweat at 3 A.M. wondering if he's right. The new electronic systems that the NWS insists on branding with wacky acronyms—NEXRAD, ASOS, AWIPS—have not yet relegated eyes, ears, and brain-generated intuitions to obsolescence. "Each place you work is a different environment," says Belville, whose career has taken him from Lubbock, Texas, to New Orleans to Washington, D.C. "It takes about a year in any given place before a meteorologist gets up to speed. You need to encounter all four seasons." *Encounter:* face scoured by the gritty Texas winds, body drenched in a squall kicking off the Gulf of Mexico. Sweat is as integral to the forecasting process as microchips. I noticed an awful lot of chewed fingernails on the hands of NWS staffers. And I found this very reassuring. After all, who are you going to trust with your life and your property—the ivory-towered theoretician or the self-described weather freak poised to jump out the window if "his" blizzard turns to rain? I don't mean to glorify these folks and God knows I do my share of railing at them whenever I don't like or trust their forecast. The NWS, like all pieces of the government bureaucracy, is cumbersome, slow-moving, hopelessly tangled in sticky red tape. For all its recent advances, the science of meteorology is still almost laughably crude in the face of the magnificent power and beauty of nature. Satellites, Doppler radar, and numerical models are like trying to illuminate a cavern by lighting a few matches. Yes. But at least the people we pay to hold the matches up are having *fun* doing it. Maybe this is the most reassuring of all.

7.

THE LONGEST-RUNNING
JOKE

THEY USUALLY SAVE IT FOR LAST OR JUST ABOUT LAST—LIKE DES-
sert or like the announcement of best picture at the Academy Awards. It
lasts anywhere from two to three minutes, depending on where you live
and what's going on that night. Over the years it has been done by seductive women in nightgowns or ball gowns, howling clowns, pigs, fish, and
puppets. It's the part of the broadcast more people watch than any other.
And they watch it as if their lives depended on it, which in some cases they
do. It's the silliest and longest-running joke on the air. What else could
"it" be but our national lullaby—the nightly television weathercast?

It's amazing how much history television weathercasting has generated in just half a century. Even more amazing is the impact of this brief
history—how radically it has altered our relationship with the elements. If
you're a weather buff of any sort, you probably can't even conceive of
weather without television. We might as well have video screens hanging

from the sky. On October 14, 1941, WNBT, the experimental forerunner to WNBC, beamed out the first televised weathercast featuring the animated cartoon character Wooly Lamb. With fewer than 6,000 sets in use nationwide, there cannot have been many viewers. But television mushroomed in the years immediately following World War II, and by 1950 there were 9.7 million television sets flickering away in the living rooms and bedrooms of America. Weather played its own bit part in the so-called golden age of television—those freewheeling pioneer days of the medium when comedians and klutzes ruled the airwaves. *Television history:* it would sound like an oxymoron were it not for the way television has made over history itself in its own image—instantaneous, loud, fast, and fast forgotten.

Considering how wild and crazy it was to become, television weather got off to a sober start. In the first few years after the war, military veterans, tweedy old professors of meteorology, and former Weather Bureau personnel barely cracked a smile as they stood in front of their wispy little maps and droned on about fronts and pressure systems. But this crew got the ax pretty fast, and once they were gone, television weather reports turned show biz. Weather was no longer just news: now it was also, and sometimes primarily, entertainment—slapstick when the weather was calm and dull; high drama when major storms hit; tragedy when the elements destroyed human life and property. Uncle Wethebee, a cartoon character who sported a handlebar mustache, helped New York's Tex Antoine chuckle out his daily forecast. In Chicago paper cutouts of Susanna South Wind battled it out with Nanook North Wind. Washington, D.C., got its forecasts from a puppet named Senator Fairweather. Television weathermen doodled, teamed up with lambs, turkeys, and gulls, and dressed up in outrageous costumes. Weather "girls" (pardon the sexist language, but that's what they were called) were all the rage for a while in the 1950s: attractive, chipper, and blessed with eternal smiles, the weather girls scrawled weather maps on Plexiglas, donned hats to match the forecast, or rose yawning from bed in skimpy lingerie to deliver the late-night forecast. Rain or shine, snowstorm or heat wave, the weather was relentlessly cute from coast to coast.

Why the glitz and shtick and antics? Tex Antoine said it laid a nice "sugarcoating on a rather dull subject," and there's something to that. The puppets and doodles and smiling women brought weather "to life" as they say—and back in the old days before satellite loops, swirling maps, and zippy color graphics, there just weren't very many lively options. Gimmicks were a way for weathercasters to "distinguish themselves from the crowd," according to Chicago weathercaster Tom Skilling.

All true enough. But I also think that television tapped into an essential absurdity at the heart of forecasting. There is the built-in risk factor that the forecast will be wrong and the forecaster will come off looking like a fool. (New York's Dr. Frank Field told me about the time his anchorperson handed him a noose on the air when the Labor Day weekend, which Field had promised would be gorgeous, turned into three straight days of chilly overcast and drizzle.) There is the widespread assumption, lingering on from the age of superstition, that the weather is by nature perverse, vindictive, irascible, or downright cruel—and that anybody who gets up and claims to understand it must be touched with these qualities as well. There is our unconscious, illogical, but nonetheless deep-seated tendency to blame the weatherman for whatever type of weather we don't want. Government forecasters can hide behind their supercomputers and numerical models—but the television weathermen are right there in our faces night after night. The clowning is thus a kind of protective coloring, a disguise that turns our suspicion into laughter. "Here is some nice guy who gets up and sticks his neck out day after day," says Allan Eustis, who worked as a television meteorologist for twenty-three years before moving over to the NWS Office of Industrial Meteorology. "The audience is sitting there just waiting for him to be wrong. Maybe it's human nature to laugh at someone who's trying to be a shaman about the weather. There's the presumptuousness of announcing, 'Here's my forecast.' And, of course, when the weather is bad people are always asking, 'Can't you *do* something about this weather?' as if we have any control. The bottom line is, it's humbling to forecast." As journalist Jay Rosen wrote in a recent essay in *Harper's* magazine, "Alone among experts on television, the weatherman is constantly being humbled by events. His outstanding trait is usually a sense of humor . . . In a sense, the weather on television is one long joke."[1] And not always in the best taste either. Tom Skilling finds the comedy tradition rather strange given the severity of America's weather: "Isn't it odd that fun and games got associated with television weather reporting and not with sports, which after all is about grown human beings playing games with balls? Sports is big business and taken very seriously, while weather—which can actually kill people—has been presented as something of a joke on the air."

Shaman or fool, the television forecaster has become a fixture of our media-driven global village. He or she has made weather not only amusing but intimate, taking highs and lows, jet streams and occluded fronts into our living rooms and bedrooms. Doing the weather can be a career path to stardom—think of David Letterman, who began his life on the air as the weathercaster for WLWI-TV in Indianapolis. Or it can be a kind of stardom in its own right—think of Willard Scott, weather prankster supreme, whose job training for the forecast slot on the *Today* show included a stint as Bozo the Clown and the creation of the character of Ronald McDonald. These people become as familiar to us as our friends. When we see them on the street we greet them by their first names. We rely on them. We judge them. We jeer them. And we just keep watching them. We really have put the weather in their hands.

How do they do it? Why do they do it? Why do they do it the way they do it?

THE ESSENCE OF TELEVISION WEATHERCASTING IS TRANSLA-tion. The television meteorologist takes the NWS products described in the previous chapter—the computer models run out of the NMC, the imagery beamed back to earth from weather satellites, the radar coverage, the zone-by-zone readings of temperature, pressure, humidity, and wind— and boils them down to a short, snappy text that will make sense to you and me. Within this basic job description there is a huge range of styles and approaches: some weathercasters really do their own forecasting, ana-lyzing, and reinterpreting the government data, pushing temperatures or

Joseph Henry, first director of the Smithsonian Institution. The network of meteorological observations and reports that Henry established in the late 1840s laid the foundation for the National Weather Service.
(*The Smithsonian Institution*)

James Pollard Espy, known in his day as the Storm King, devised and promoted a highly influential theory of convective storms. His 1841 volume *Philosophy of Storms* was required reading for nineteenth-century weather buffs.
(*The Smithsonian Institution*)

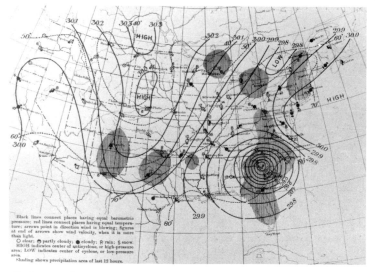

Early map of West Indian hurricane, August 28, 1893, 8 A.M. By this time the Signal Service had evolved into the United States Weather Bureau under the auspices of the Department of Agriculture.
(*National Archives*)

Street scene in New York City after the blizzard of 1888. The official "indications" called for rain followed by "fresh to brisk westerly winds, fair weather"—one of the most notorious forecast bloopers in Weather Service history. *(Library of Congress)*

An estimated six thousand people died in the hurricane that devastated Galveston, Texas, on September 8, 1900. The Weather Bureau warned of the approaching storm but failed to predict the impact of the high tide. *(Library of Congress)*

The historic Trinity Episcopal Church in Pass Christian, Mississippi, before and after Hurricane Camille ripped through the Gulf Coast on August 17, 1969. The storm surge created by the storm pushed Gulf waters twenty-four feet above their normal level. *(NOAA)*

Measuring snowfall the old-fashioned way on the roof of the central office of the old Weather Bureau in Washington, D.C. Under the current National Weather Service modernization program, automated weather instruments are becoming standard at weather stations and forecast offices around the country.
(National Archives)

A hurricane as it looked on one of the first weather radar sets. Used widely for military purposes during World War II, radar became an important tool for local forecasting after the war. The NWS modernization program is currently replacing all conventional weather radar with the more advanced Doppler radar.
(National Archives)

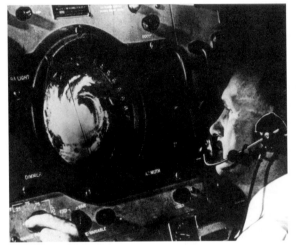

Last-minute adjustments made to the TIROS (Television Infrared Observation Satellite) prior to its launch in 1960. The satellite orbited the poles at about five hundred miles up, beaming back pictures of earth's cloud cover taken by two television cameras.
(National Archives)

First image from GOES-7, the National Oceanic
and Atmospheric Administration's seventh
Geostationary Operational Environmental
Satellite, which was launched in February 1987.
Since their orbits keep exact pace with the
rotation of the earth, the GOES satellites
in effect hover over a fixed point on the earth's
equator at about twenty-three thousand miles up.
(NOAA)

Satellite view of Hurricane Hugo,
which did extensive damage
to Charleston, South Carolina,
and surrounding areas in September
1989. Predicting the path of a
hurricane is notoriously difficult,
even with satellite imagery,
but the National Hurricane Center is
getting better at it.
(NOAA)

Technicians in Titusville, Florida, unpack the
GOES-J, the latest addition to the family of
Geostationary Operational Environmental Satellites.
This photo was taken in February 1995,
three months before the satellite was due to be put
into orbit aboard an Atlas I expendable launch vehicle.
(NASA)

Weather balloon readied for launch.
Instrument-bearing weather balloons are
released twice each day all over the world,
and the data they record is a key element in the
computer-generated numerical forecast models.
(National Archives)

The dawn of numerical forecast modeling, way back in 1962. In those days, clerks entered weather data on punch cards and then fed them into an IBM STRETCH computer. After digesting the data, the computer churned out crude "weather flow pattern forecasts," shown at right.
(NOAA)

Computers come of age. By 1974, when this photo was taken, magnetic-tape drives had replaced the old punch-card system, and the IBM 360/195 was the computer of choice at the National Meteorological Center in Suitland, Maryland.
(NOAA)

MRF CONDITIONAL PROBABILITY OF SNOW (dashed)
12-H PERIOD ENDING 0000 UTC 14 MARCH 1993

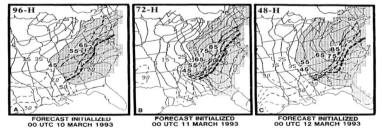

FIG. 4. The 5-, 4-, and 3-day conditional probability of snow (CPOS), derived from the MRF, valid for the 12-h period ending at 0000 UTC 14 March 1993. Values greater than 45% are shaded.

The "Superstorm" of March 12 through 14, 1993, as foretold by the Medium-Range Forecast Model, the workhorse global model used by the National Meteorological Center. These maps, showing the "conditional probability of snow," were generated five, four, and three days before the event.
(National Meteorological Center)

A very young Tex Antoine, one of the first and best-loved television weathercasters, concocting his forecast back in the 1950s with the help of Uncle Wethebee. Note the rudimentary weather map and the logo of WNBT, a forerunner of WNBC. *(The Bettmann Archive)*

Vivian Brown, one of the twenty-four on-camera weathercasters who bring you accurate and dependable weather forecasts twenty-four hours a day—only on the Weather Channel. Since its debut on cable television in 1982, TWC has revolutionized television weather. *(Courtesy of the Weather Channel)*

Storm-driven waves. Some climatologists believe that global warming will intensify the frequency and severity of tropical storms, though this idea remains open to debate. *(NOAA)*

Flooded river town. The catastrophic floods that hit the Mississippi region during the summer of 1993 and the state of California during the winter of 1994–95 have been linked to ENSO, the El Niño Southern Oscillation that shifts global weather patterns every four to seven years. Scientists speculate that ENSO events may become more common in a greenhouse world. *(NSW)*

precipitation likelihoods up or down on the basis of their own calculations, hunches, and weather wisdom; others are more or less actors who take the NWS guidance and deliver it with a smile; still others hire private weather companies to do the analysis for them. But no matter what their approach to forecasting is, they all get their baseline data from the same source. The familiar tag line "Here's my exclusive Channel Q forecast" is a lie. "I hate this 'here's your forecast' business," says Eustis. "We all look at the same maps, the same numbers, the same progs. Yes, there can be different interpretations, but 90 percent of the time all the stations are so close that you wouldn't change your clothes or your activities no matter what station you watch." As Ray Boylan, a weathercaster on WSOC-TV in Charlotte, North Carolina, puts it, "We are a delivery system for the NWS." All television weather is a set of variations on a theme. And that theme is written anew each day by the NWS.

Television weathercasting is also a business, a highly competitive business with a fierce dedication to the bottom line. Accuracy doesn't matter a whit if your ratings slip. Entertainment has always been the name of the game on television—and the power to define what's entertaining and what's not has always been in the hands of a notoriously skittish, fickle management. "Make it folksy," these demigods of the airwaves command at the start of one fiscal year—and weathercasters from coast to coast sprout patches on their elbows and chuckle about their gardens and the sweaters their wives are knitting. "Get rid of these hayseeds, and put a weather scientist on," comes the next directive a few years later, and suddenly the airwaves are buzzing with lectures on convective storms and 500-millibar height patterns. Private consultants then inform management (for a princely fee) that ratings would leap if the weather report was more dramatic—and overnight every passing shower becomes a "precipitation event" and each gust of wind a "potentially life-threatening situation." The needle has been swinging back and forth between silliness and science for nearly fifty years now and shows no sign of coming to rest.

Silliness reached its absolute zenith back in the 1950s, the age of the weather girls. Nobody back then gave a hoot what the weathercaster said, so long as she had nice legs. "The weather report was a place where you could stick in a pretty face," one (male) veteran of those old days told me, "and a nice body didn't hurt either. It was comic relief. But the truth is, these weather girls were not skilled. None of them knew what the hell they were talking about—and they didn't care. They just read whatever came off the weather teletype. They'd take a Weather Bureau map off the fax, copy it, maybe put in a smiling sun or a cold front now and then. That was it."

Carol Reed was the leader of the weather girl pack. Probably the first woman to do weather on television, Reed debuted on the air in 1952 at age twenty-six with no qualifications aside from a cheerful manner and knack for communication. For twelve years she faced the cameras at New York's WCBS, smiled brightly, zipped through the forecast, and signed off each weathercast by telling her viewers to "Have a happy." Her reign ended abruptly in 1964, when the station put her out to pasture. Robert Henson, author of *Television Weathercasting,* reports that after Carol Reed's success, stations all over the country wanted to have weather girls of their very own. By 1955, according to Henson, "women made up a major fraction of all weather anchors, if not a majority."[2] The New York *Times* estimated in 1961 that as many as three-quarters of America's 466 television stations employed a weather girl. Washington, D.C., viewers could choose between "USO singer-dancer-comedienne" Cindy Dahl and Tippy Stringer with her cute little cartoon character called Senator Fairweather; Kay Field, a "former name-band singer" was on the air on WISH in Indianapolis; Jan Crockett strummed the ukulele on New York City's Channel 7 and gushed that she "simply adores weather"; Eugenia Burke, whose credentials included the title "Miss Press Photographer," smiled her way through rain or sun at WARM in Scranton, Pennsylvania. Tedi Thurman, whom *TV Guide* dubbed "weather girl supreme," turned her regular weather spot on Jack Paar's *Tonight* show into something akin to a striptease. Forecasting from behind a shower curtain, she poured on the sexual innuendo: "In the morning—rain," she would purr. "In the afternoon—dew. In the evening —don't." "Virginia will be very warm, and I'm just thrilled for poor Virginia."[3] One contemporary observer dismissed the whole race of weather girls as "fashion plates with a *touch* of weather." Many stations hired backstage (male) meteorologists to brief the women and write scripts that they recited from memory on the air. *The New York Times Magazine* reported in a story called "The Weather Girls Ride Out a Storm" that Carol Reed was actually experienced enough to make her own forecasts but that CBS management would never permit this. It was the golden age of television and the last hurrah of pure, unadulterated sexism. Today Willard Scott carries on the weather girl tradition—not so much for his sex appeal but for his unquenchable silliness and frank meteorological ignorance. "The forecasts Willard gives are made directly by the NWS and that's fine with him," one of his colleagues told me. "That way, they get the blame if the forecast is wrong. He's making millions—he doesn't want the credit." As Scott himself once put it, "A trained gorilla could do this job."

Gradually, during the late 1950s and early 1960s, the weather girls were replaced by trained professional meteorologists—Don Kent in Boston, Harry Volkman in Chicago, Harold Taft in Dallas, Jack Capell in Portland Oregon, Louis Allen in Washington, D.C., Francis Davis in Philadelphia, and Stormy Rottman were some of the more celebrated pioneers. Robert Carnahan, now the commissioner for professional affairs of the American Meteorological Society (AMS), told me about his first television weathercasting job in Connecticut in 1955: "Back in those days there was no set, no satellite imagery, and no computer graphics. Most of our information came from the old Weather Bureau via teletype and I would do the maps myself. I used paper maps and stuck pins in them to show the fronts. The next day's forecast was as far out as we'd go—thirty-six hours was the limit." Another old-timer recalls that the Weather Bureau did issue a general five-day outlook back in the 1950s, but "it was so bad you were crazy to use it. It was like shooting darts at a dartboard."

Bob Copeland went on the air at Boston's WBZ in 1957, soon after getting a degree in meteorology from MIT. He knew a lot of weather science, but had absolutely no broadcast experience. "I was so nervous on those first shows I was convinced I'd have the shortest weathercasting career in television history," Copeland remembers today. "I had diarrhea for two months." But as it turned out, he stayed on the air in Boston for the next thirty-six years (he moved to a different station in 1966), setting some kind of weathercasting record when he finally retired in 1993. He laughed when I asked him to describe the state of the art of weathercasting back in 1957: "We had big maps made out of slate and we wrote on them with chalk. The maps were heavy and they were set up with window weights so you could push one up to reveal the one behind it. Every now and then there would be a disaster: the ropes would break and the map would come crashing down and there was no getting it up again. I almost had one of my hands guillotined off by those maps. Later on, we went to paper maps thumbtacked to a revolving panel. That was a bit of an improvement, except when the map fell off and landed on the floor. All of this, of course, happened live on the air." Back in the old days, Copeland had a full five minutes to do the weather on a half-hour news broadcast—an eternity by current standards.

"It used to be fun and games," reminisces New York's Dr. Frank Field, who has been at it since the early 1960s and is still going strong. "In fact, it still is at some stations. They'd have a chimp or a ukulele player. Nobody cared. It was nobody's weather. I remember when I started on television some thirty years ago, there was an instant when we lost a hurricane. We had no idea where it was." "There was a general perception that forecasts were not very good back then," Carnahan admits, "and in fact they weren't." What television people call "production values" were pretty much nonexistent back in the 1950s. The more creative weathercasters would doodle rapidly to show the movements of fronts or stand behind clear Plexiglas screens and draw in the highs and lows and wind directions on the reverse side, so that the viewers could see both the map and the mapmaker through the map. But that was about as animated as things got. Remember, too, the state of the art of the television set in those days: a tiny black-and-white screen on which bands of "snow" rose or fell regardless of the season. No wonder they brought on the chimps and puppets and cartoon characters.

Some weathercasters managed to rise brilliantly beyond these limitations. Eustis singles out Charles Mitchell as one of the all-time great forecasters from the old days in the Washington, D.C., area: "He came up with a long-range forecast, out to seventy-two hours, which was a big stretch back then. And he did it largely on his own, from memory and experience. He would say, 'I've been looking at pressure falling in Grand Forks and winds in southern Canada, and when I've seen this before, we'd get rain in seventy-two hours.' He had a wealth of knowledge stored up from pattern recognition. He didn't need mathematical models—he worked on hunches and he put things together that no one else was looking at. He was one of a kind."

THE PROFESSIONAL METEOROLOGISTS ACHIEVED SOME OFFICIAL recognition when the AMS inaugurated its "seal of approval" program in 1959: the seal was awarded to meteorologists who conducted themselves on the air with dignity and seriousness and who succeeded in making weathercasts both informative and interesting. The seal program helped sober up weathercasts during the course of the 1960s, though it did not stamp out buffoonery altogether. During the 1970s there was yet another shift in television weather when the "happy news" format caught on. Now that he or she was a member of the grinning family of television journalists known as the "news team," the weathercaster was supposed to "interact," make pleasant chitchat, kid around, and address the other anchors fondly by first name before delving into the weather report. The weather, being what it is, inevitably inspired lame jokes or anguished laments, and weathercasters fell naturally into the role of class clown or family idiot. The requisite "happy news" banter also sheared as much as a minute off the precious two to four minutes allotted to weather. The new chummy format gave the weather report a different "feel," but through much of the 1970s, the content of the television weather report remained substantially the same as it had been from the beginning of the medium: the same static hand-drawn maps, the same dull graphics, the same inevitable progression through readings, national picture, local forecast, long-range outlook.

New technology and the proliferation of private weather companies revolutionized television weather in the late 1970s. A few pioneers, among them George Winterling at WJXT in Jacksonville, Florida, and Dr. Frank Field, then with WNBC in New York, had experimented with putting satellite imagery on the air as early as the 1960s. But these were for the most part fuzzy still photos or laboriously contrived and jerky animations. The goal that no one had yet attained was to put the maps in motion: to get images of weather systems to slide and billow across television screens just as they do in the sky. The 1975 launch of GOES-1, the first geostationary satellite, was a crucial breakthrough in animating the weather segment: since GOES beams back an image every half hour from a fixed point over the earth, weathercasters could easily splice together a series of shots showing cloud movement over the entire continent in the course of a day. These cloud cartoons, known as satellite loops, are now a standard feature of most weathercasts. Radar imagery also improved rapidly in the course of the decade, with new techniques devised to display intensity levels within storm cells and color added by the end of the 1970s.

The personal computer, which spread through the world like an electronic epidemic during the 1980s, coordinated and enhanced all these new visual elements, making weathercasts zippier and brighter than ever. Ingenious start-up companies like Weather Central in Madison, Wisconsin, designed integrated hardware and software systems that took NWS data and converted it into full-color animated weather maps. Overnight, the computer rendered obsolete the old-fashioned maps with their pins, chalk, markers, and Plexiglas. "The new technology came along and suddenly television weather became a different ball game," says Frank Field. "If you didn't have the latest development, you looked like an idiot."

One piece of computerized high-tech that everyone in the weather business jumped on was chromakey. In the chromakey process, the image of a weathercaster pointing to a weather map and the image of the map itself are filmed separately and then electronically merged and transmitted together. In the television studio, the weathercaster actually stands in front of a blank, uniformly colored screen (usually green or blue) and points at a void; a computer overlays the weather map into the "hole" so that a single seamless image appears on our television screens at home. Since the weather map really isn't there behind the weathercaster, how does he or she know where to point and when? The trick with chromakey is that the map is displayed on monitors placed to the side or above the camera. The weathercaster has to learn how to coordinate and time his or her gestures at the chromakey screen to match up with the map (or series of maps), and to do it without squinting too obviously at the monitor. It's a bit of sleight of hand, like pulling the old bunny out of the hat.

As the computer revolution opened up new opportunities in weather communications, private weather companies came rushing in with creative new ways to package, apply, and market them. Some of these companies have actually been around since World War II, preparing and selling specialized forecasts to industry. But this sector of the weather business really took off after the computer revolution. Operations like WSI, Kavouras, and Accu-Weather have in effect made themselves into high-tech commercial clearinghouses for weather information: they take the weather products generated by the NWS; refine, reformat, reconfigure, recolor, and retool them; and then sell the enhanced weather packages to clients ranging from commercial television stations to air carriers, newspapers, utilities, theme park operators, and agribusiness managers. When you tune in to your local weathercast, chances are that at least part of the show has been created by a private weather company: it might be the graphics or the maps or the satellite loops, it might be the forecast itself (Accu-Weather does this), or it might be some more subtle and less visible element, like the format of the numerical models that the television meteorologist consults in making his or her predictions.

Bill Schlueter of the Minneapolis-based Kavouras Inc. describes the basics of how his company services television stations: "When Steve Kavouras started the company in 1976, our first effort was to convert black-and-white Weather Service radar into color video imagery. We have become famous for the high resolution look of our weather imagery: we've totally gotten away from that blocky, chunky computer-generated look. In the area of television, we provide the stations with a constant stream of data. When we started out we communicated with our clients over the phone lines, by fax and modem, but this has a built-in limitation—it's slow and very expensive. So in 1982–83 we got into the satellite transmission business and now we transmit data right off a communications satellite. We send stations geostationary satellite imagery that we get from both NOAA and European sources; we make satellite animations; we have products that enable you to fly around and through the weather; we generate forecast graphics—the positions of fronts, high and low pressure; we can get a radar image of rain falling anywhere in America in as little as two minutes after the event; we even manufacture software that enables TV meteorologists to create their own animations from raw data—they plug in the numbers and the program spins out a map displaying time-lapse sequences, temperature readings, or whatever else they want. We diddle with the data so that it works better for the individual client."

Tom Skilling, a weathercaster on Chicago's WGN, is one of the clients who pay for the data that Kavouras "diddles with." Here's what the partnership looks like from his end: "We have a 280,000-byte data feed from Kavouras and we get the full suite of global and hemispheric models from them. Kavouras takes the avalanche of data from Washington, D.C., and gets it to us in a timely fashion. They take the raw grid points off the models and plug them into a mapping program and then they send it to us in a form we can use. The NWS cannot go through that level of post-processing. We get our radar and satellite imagery from Kavouras, but I do my own graphics. Companies like Kavouras have revolutionized media weather even for small television stations."

Private weather companies have moved aggressively (and profitably) into areas of forecasting that the NWS cannot or will not handle: "customizing" weather data to suit the individual needs of clients, pinpointing storm paths at a specific crossroads or city block (as opposed to the zone or county forecasts that the NWS issues), providing slick, bright, user-friendly maps and readouts. Most of the people I spoke to (on both sides) described relations between the government and the private companies as friendly and cooperative, in part because their responsibilities and functions are so different. "The private companies are a $200-million-a-year business," says Eustis, "and the government can't possibly take over that work. The public/private partnership is a good thing." "We need them a lot more than they need us," admits Schlueter of Kavouras. But friendly rivalries are inevitable. As Mike Smith, president of a private weather company called WeatherData Incorporated, based in Wichita, Kansas, puts it, "If you had two sportscasters trying to predict who would win this weekend's NFL game, of course there would be some rivalry. Everyone wants to be the best." Income discrepancies breed rivalries of a different sort: private meteorologists stand to make considerably more money than NWS staffers, whose salaries are fixed by the civil service pay scale. More serious tensions have arisen, however, when private companies trespass on territory that the government considers its exclusive domain: the issuing of severe weather warnings. Although there is no law stating that storm warnings may only be issued by the NWS, it is part of the explicit policy of the NWS, and everyone in the weather business acknowledges that this is the way things work. But not all meteorologists can restrain themselves from jumping in. Accu-Weather raised NWS hackles a few years back when Hurricane Gilbert was taking aim at the city of Galveston. On the basis of Accu-Weather warnings, parts of the city were evacuated; the National Hurricane Center, however, had issued no warning, predicting that the storm would pass south of the city, which is in fact what happened. Accu-Weather justifies its action as freedom of speech. But the NWS has not forgotten or forgiven. To this day, the name Accu-Weather brings a pained look to the faces of NWS personnel.

ABOUT HALF OF THE WEATHERCASTERS IN THE TOP FIFTY TELEvision markets today are professional meteorologists, with some sort of degree in the science of weather. The rest of the pack runs the gamut from amateur weather nuts to bright smiling pretty faces who have mastered the weather patter to slick communicators with shellacked hair to out-and-out clowns who don't know an isobar from a candy bar. But no matter what their relationship with the weather is, all weathercasters must be actors. They stand up there day after day and broadcast a fixed and instantly recognizable image—a made-for-television persona that is as crucial to their forecast as the current temperature and air pressure. Even if their predictions depart not a jot from the NWS guidance, they need to make us *believe* that it's "exclusive," "unique," and of course more accurate than any other show in town. How much we believe is the heart of the matter. The whole game hinges on our loyalty: whoever gets the most viewers to stay tuned for the longest amount of time wins. When you get right down to it, television weathercasting is really about the personalities of the weather actors: the attitudes and feelings they project about weather, the jokes they tell, their enthusiasm, their choice of words, the bits of science and lore they toss in.

I asked a few television meteorologists to step back and talk about this aspect of their work. Here are some of the verbal snapshots I came up with:

DR. FRANK FIELD, WCBS-TV, NEW YORK CITY

Dr. Frank Field (not a medical doctor, as everyone assumes, but a doctorate in optometry) holds the title of New York City's longest-running TV weathercaster. After serving as a weather officer for the air force during World War II, Field went to work for the Weather Bureau in the 1940s and 1950s and then signed on with WNBC in 1957. In 1984 he moved over to WCBS. Field's kindly, homey, avuncular image is enhanced by having his son Storm Field share the forecasting duties with him at CBS. (Storm had previously been at WABC for fifteen years, so before father and son became partners at Channel 2 they were in nightly competition. "Mother used to tell us who was right," Field says, laughing.)

"I like to explain things on the air in an elementary way without being overly complicated. Of course, you also have to have a sense of humor. Back in the old days at NBC, Johnny Carson and I got into this joking feud. Carson taped his show at 5 P.M. and we were on at 6:30. People started telling me about the nasty comments he was making about me. He started calling me 'NBC's crack meteorologist.' He said, 'Field was the lookout at Pearl Harbor.' So the next night I got an outtake of Carson making a twisted face and I had the station artist paste this into the middle of a group of Japanese dignitaries looking at the cherry trees in Washington. Then Carson invited me to come on his show, and they dumped a bucket of water on me. It was the biggest laugh I'd had in years. You've got to roll with it. Television, after all, is entertainment.

"I got hired at NBC in 1957 because management was upset over the fact that there were no real weathermen on the air in New York City. This was the era of Tex Antoine and Carol Reed—both of whom I taught meteorology to. When I first went on they paid me seventy-five dollars for each show and I thought this was a hell of a lot of money. I didn't even know which camera to talk into. I fumbled my way through. The truth is, television back then was not brain surgery—and it still isn't to a great extent.

"The TV weatherman is typecast as a nice guy, somebody who is fun to talk to. People come up to me on the street all the time. 'Hey, Doc,' they call out, 'what kind of weekend will we have?' That kind of thing. Nobody would talk to Walter Cronkite that way. Weather is an equalizer —it's an everyday kind of subject, something we all experience together. You get heat from management to make it nice.

"Weather forecasting is a science, but it's also a seat-of-your-pants kind of endeavor and an art too. We all work from the same readings, but not everyone agrees on how to interpret them. And then it's impossible to take into account all the different microclimates in the metropolitan area. The south shore of Long Island is a world away from the city in terms of weather. But even in Manhattan, when I walk from midtown to the West Side I'm aware of a cooling breeze off the river. There are so many variables.

"We're also limited by how short a time we have on the air. We can't properly do *local* weather in two minutes, let alone national. Sometimes we only have one minute.

"We don't keep track of our accuracy rates and compare our scores with the other stations. That is childish. We're all using the same materials. Sooner or later you'll be right and they'll be wrong and vice versa. When you're wrong you don't want to commit suicide. It's really not that big a deal."

RAY BOYLAN, WSOC-TV, CHARLOTTE, NORTH CAROLINA

Boylan spent twenty-four years doing meteorology with the navy before launching a career in commercial television in 1975. His first job was in Jacksonville, Florida, and after eleven years there he moved on to Charlotte, North Carolina, where he has done the weather on WSOC-TV since 1986.

"The real challenge of my job is to make people want to watch me without being patronizing or silly. In this business, success is measured in ratings—and you always must keep an eye on the bottom line. Let's face it, the viewers have three options—me, the weathercaster on WBT, and the one on WCNC. I want my product to be the best. If you were to beat me with a bicycle chain, I'd admit that I devote about 15 to 20 percent of my energy to meteorology and 80 to 85 percent to production. The show has to look good.

"Have you noticed that weather is rarely placed early in the newscast? We put it deep in the show to keep viewership. Weather is the major draw in local news. Most people won't stop watching the news until they know the weather. The attitude is, 'I have to find out what it's going to be tomorrow so I can go to bed.'

"The day I started on television, I walked into the station in Jacksonville at 3 P.M. and said, 'I'm here, what kind of training should I do?' 'Anything you can get in before 5 P.M.,' they told me. I was *awful!* In those days we were still using colored pencils on a map with an art card. I suffered through the first few weeks.

"These days, management wants people to be attractive and informative from the get-go. We're in the era of consultants who are hired to cook up new things to improve ratings. I'm on the receiving end of this. The last thing I heard is they think I should grow a garden. But I hate gardening. Whatever these consultants want is going to happen. If you are a weathercaster, you are a willow and you will bend.

"Communication is the main thing. You're going to confuse people if you drone on and on about strong upper-level disturbances. On the other hand, people have become much more informed thanks to television weather. These days people are more attuned to the threat of killer weather. In New York City they might have forty-five seconds of a clown —but outside of the major metropolitan areas there has been a dramatic increase in public awareness."

TOM SKILLING, WGN, CHICAGO

A self-described weather fanatic since early childhood, Skilling began making his own forecasts off the NWS daily weather maps when he was still in elementary school. At the age of fourteen he struck a deal with his local radio station, WKKD in Aurora, Illinois: if they would special-deliver to him the current daily weather map, he would do the local forecast. The station loved the idea of a kid doing the forecast, and they kept him on for three years. Skilling has been in the weather business ever since. He did radio weather during his college years at the University of Wisconsin, moved on to television in Milwaukee, and in 1978 he landed his current job at Chicago's WGN. Though Skilling has been weathercasting for the better part of three decades, he still has the same boyish enthusiasm for atmospheric events.

"When I was back in Milwaukee I worked with a puppet for three years. This puppet had been on television for twelve years running and was super popular—they just couldn't drop him from the show. Gimmicks and props were big on weathercasts back then because you had to have *something* to distinguish yourself from the crowd. But these days I don't have to stand on my head—I do it with computer-generated graphics. We've reached a new level with fly-through sequences in three dimensions. The high-tech graphics free me to do straight interpretation. I try to give viewers a sense of how the atmosphere is supposed to evolve. I show them the approach of a cold spell or heat wave by pointing out the jet stream. In the winter I'll point out how the jet stream comes crashing down on the Midwest with a load of arctic air.

"I have three and a half minutes to do the weather on our noon show and again on our 9 P.M. show. I bask in the glory of those three and a half minutes, because these days this amount of time for the weather is almost unheard of.

"I try to make sure all my viewers understand everything I say. Whenever I give a forecast out to seven days, I always qualify it by pointing out that the last few days are not as accurate as days one and two. But public perception of our accuracy still lags seriously behind reality. Most people have no idea how we derive the forecast. The attitude that 'they're wrong all the time' persists. People think we're in here throwing darts at a dartboard. So I try to give people insights into how meteorology is done today.

"I do think there is a widespread and increasing interest in weather these days. People love the detail. They are interested in how weather affects them physically and psychologically.

"I remember how the first numerical models coming out of the NMC were so hard to read. The early models were bad at picking up convective storms—you had to add the thunderstorms in. Today, instead of having two primitive models, we have six advanced ones that we can compare with each other and attempt to reconcile. Yet I feel that numerical modeling has led to a degree of laziness. We meteorologists are supposed to add quality and resolve any bias in the models, but some people take the model guidance raw and run with it.

"The forecasts we do on the air are not one person's forecast—we all lean on the work of other people. Our trademark here is that we spend a lot of time putting it all together. If you can improve on the model guidance, more power to you. A reputation for getting it right will distinguish between you and the other forecasters."

Pam Daale, Chief Meteorologist at KMGH-TV, Denver, Colorado

Pam Daale began her broadcasting career in 1985 when she was still in college at Texas A&M. She landed another weathercasting job the following year when she transferred to Iowa State, and she has worked in TV meteorology ever since. Physically disabled, Daale delivers her twice-daily forecasts from a wheelchair. But she says she feels even prouder of being a woman meteorologist (as opposed to a "weather girl" with no professional background in meteorology) than she does about working around her disability. "I believe I am the first female meteorologist on the air. I feel a lot of pride in my training as a scientist."

"I grew up on a Midwest farm and I know how important weather is to people's livelihoods. Partly as a result of this, I am very matter-of-fact in my forecasts. We are having a drought here in Colorado right now, so when it's 80 degrees and sunny, I don't say, 'Gee, isn't it terrific that we're having such nice weather.' I mention that we could sure use the rain. And I steer clear of comedy altogether.

"I like to say that my forecasts are 100 percent me and 100 percent the NWS. All of the input comes from the NWS—all the models, all the numerical guidance, and the satellite imagery. But I go over all the data myself. I crunch it all through my own brain.

"We subscribe to WSI [one of the big private weather companies] and our satellite loops and radar imagery come from them, but we make our own base maps and put our own information on top of them. The computer is continually ingesting data, and I call up what I want to look at.

"My first show is at 5 P.M. and my typical day starts at 1:30 P.M. I read through the information on the weather wire, which we get from WSI, and if it's raining I look to see if it's intense. I familiarize myself with the maps. I look at the satellite loops, the water vapor loops, and then I look at the closer-in data—the region's temperatures, winds, and dew points. I write down my forecast on paper and then transfer it to a computer before the 5 P.M. show. My crunch time is between 4:30 and 5. Then I break for dinner, catch up with whatever is on my desk. Then I'm on the air again for the 10 P.M. news.

"Of course, all of this is different when we're having severe weather. Only yesterday we had conditions for a tornado watch for the eastern third of Colorado. When this happens you have to forecast more quickly, you have to get the warnings on the screen instantly. You're glued to the radar trying to predict which cell will pop next. The tornado warnings come directly from the NWS. I never would issue a warning independently, though I might say strong storms are heading for the Parker area. The NWS is so sophisticated these days—they know a lot more than I do.

"I don't really think of myself as a descendant of the weather girls because they were not meteorologists. But I do have a lot of respect for the ones who took a real interest in the field and learned about meteorology. This is a male-dominated field and I give them a lot of credit.

"All people who appear on television are placed on a pedestal by viewers. And when you do the weather, people have a kind of chummy attitude toward you. Weather is one thing that affects every single person. Even if they're not up on current events, they see weather and live with it every day. So they can really relate to the weathercaster.

"Television is entertainment. That's good and bad. The funnier someone is, the more people will watch. But I don't think of myself as an entertainer. I love weather. I'm much more into the meteorology than I am into the television part of it."

BOB RYAN, WRC-TV, WASHINGTON, D.C.

Ryan grew up in New York State's Hudson Valley region during the Tex Antoine era and went on to get his BA in physics and MS in atmospheric science from the State University of New York at Albany. After six years of weathercasting on WCVB in Boston, Ryan moved on to the *Today* show in 1978 and then in 1980 to WRC in Washington (he and Willard Scott swapped positions). A former president of the American Meteorological Society, Ryan has earned the respect of his colleagues in the NWS as well as the loyalty of a large segment of the Capital District viewership.

"My basic feeling about the weather is that there is always something interesting going on. When I'm on the air I'm always enthusiastic about the weather and I always point out the positive side of things. There is nothing wrong with a rainy day, or bitter cold, or a heat wave. It's not the end of the world. Life goes on. I don't grumble about it. I share a bit of the science of meteorology with viewers and communicate my enthusiasm for what lies behind the weather.

"There is a delicate balance between being entertaining and being instructive. Here in Washington we have many viewers with science or engineering degrees, so I can go into a little more depth about the explanations for weather events. I tie it in to what's going on that day—be it hail or a shift in the jet stream or something of that nature. Of course, you don't want to make it Meteorology 101. You try to match the interests of your viewers. For example, in Florida people always want to know about the weather back home in Minneapolis or Boston. So that becomes a part of the weathercasts there.

"What's really important in weathercasting is how the information is communicated. I may have a highly accurate forecast, but I must communicate it in an understandable way so that people will retain it. A lot of it comes down to your choice of words. For example, the NWS forecast might say 'sunny and pleasant' but I'll say something like 'azure-blue sky, low humidity, with puffy cumulus clouds in the afternoon.' I give a picture of the day because people remember this.

"There are times when I disagree with the NWS forecast—for example, they might be predicting six inches of snow followed by rain but I'll conclude from the models that only two inches of snow will fall before it changes over to rain. But day in, day out the differences are not that great. It's like going to two or three different doctors—the symptoms are the same, but each doctor may come up with a slightly different diagnosis.

"The event that really changed television weathercasting was the launch of the TIROS satellite in 1960. Now you could actually *show* the weather—it was a tremendous tool for forecasting and a tremendous television tool. Since then, people have become more and more interested in what will happen in the weather, especially with dangerous weather.

"These days it's important for the credibility of a station to have weather people who seem to know what they're talking about. You don't necessarily have to be a professional meteorologist to communicate weather information. If you work hard and are diligent and try to be thorough, I have no problem with that."

GARY SHORE, KJRH, TULSA, OKLAHOMA

"Oklahoma is the most volatile place for weather in the country," asserts Gary Shore, and in his eleven years of weathercasting at KJRH, he has received high marks for tracking this volatile—and often deadly—weather with impressive accuracy. KJRH installed Doppler radar in 1984, at a time when the National Severe Storms Laboratory in Norman was still experimenting with it, and in the past decade Shore has become expert in using Doppler to stay on top of tornadoes and other severe storms. Shore got his undergraduate degree from New York University and received his master's in meteorology from Penn State University. His career in weathercasting has included television jobs in Wichita, Kansas, and Washington, D.C. But Shore prefers working in Tulsa because, as he puts it, "people really care about the weather here."

"Tulsa has been declared a disaster area more than any other locale. The extremes here are incredible. On June 8, 1974, we had flash floods after receiving ten inches of rain. In the summer of 1980 we had fifty-eight days over 100 degrees and almost no rain. Four years later, fourteen inches of rain fell in five hours, killing fourteen or fifteen people and causing $300 million of damage. Flash floods coming after a major tornado outbreak just about wiped out the town of Manfred. In April 1993, we used the Doppler to see how a storm was organizing itself for a tornado. I went on the air an hour before it hit to alert viewers to be ready for a possible tornado. The official warnings came out only two to three minutes before it hit the ground. Then, over a minute before, we saw the actual air spinning up like the eye of a hurricane and I jumped on the air and told people block by block where the tornado was heading.

"The NWS has the ultimate responsibility for issuing warnings, but when I see a tornado forming or when one of our spotters calls in with reports of conditions favorable for tornadoes, I am obligated to go on the air. I call it a nonwarning warning. When dangerous storms threaten, we take over the airwaves hour after hour. We are nowcasting. Last week we were on for two solid hours when we had four inches of rainfall in an hour.

"In my regular weather broadcasts I try to explain the weather in ways that are understandable to all viewers. I relate the weather to the real world. My show is fast-paced and straightforward—oriented toward information, not folksiness. I am known for accuracy, so this is where I put most of my energy. People appreciate that here. People come up to me and say, 'Hey, Gary, you get that weather pretty darn close.'

"There are basically two styles in weathercasting—the showman who is a gifted storyteller and entertainer and the professional meteorologist. I strive to be both. I strive to be warm and friendly and creative, and also honest. After all, most people watch the news for the weather. I have an opportunity to brighten their days—to make them feel that there's a friend out there who is interested in the same things they are. A big part of my credibility is that I am honest with people. I also hope my enthusiasm for weather comes through.

"Weather forecasting is like solving a puzzle. I'm sitting here right now with 120 weather maps, hundreds of satellite shots, radar imagery. There's more data than any one human being can look at. My accuracy ratings are around 90 percent or in the low 90s. It's very rare that we run into something totally unexpected—maybe one day out of twenty.

"One of the secrets I've learned is to avoid holding on to your last forecast out of pride. The atmosphere is volatile and fast-changing, and you need to be prepared to flush away that old forecast and start over. Each time you forecast you should look at things with fresh eyes.

"I was always into weather forecasting. Even as a kid, I would run in from stickball to watch Tex Antoine. Then when I was a graduate student at Penn State University I got to work on television. My first time on the air I was like a deer in the headlights. Really scared. The general manager of the station said, 'There's no way—figure out something else to do, son.' But I got better slowly.

"One of the most interesting things I've ever done as a meteorologist was to chase tornadoes. I stood one mile from an F5 [the most violent category on the Fujita-Pearson Tornado Intensity Scale, with winds of 261 to 308 miles per hour] and it sucked the glasses right off my face. We got twenty minutes of continuous coverage of that storm."

"WHEN YOU'RE A WEATHERCASTER YOU'RE BEING INVITED INTO someone's living room or bedroom," Allan Eustis remarked as he looked back on his own twenty-three years in the business. "You become their friend. You also get power. People think you can control the weather, like an Indian sage. You're forecasting the domain of the gods."

But is it possible that the intimate, illusory friendship between weathercaster and viewer is destined for obsolescence? John Coleman certainly thought so when he started the Weather Channel (TWC) in 1982. All weather, all the time: if you could flick on TWC whenever you felt like it and get the whole ball of wax—national radar and satellite view, tropical update, five-day national forecast, international report, and of course your local forecast—why hang around waiting for Willard Scott or Bob Ryan to do their two-minute routine fifty-five minutes into the show? Coleman knew firsthand how confining that routine could be: he had put in five years as the weathercaster on *Good Morning America* as well as working at stations in several major cities in the Midwest. Forecasting a continent's worth of weather in two and a half minutes—and often less when something glitzier preempted the weather slot on *Good Morning America*—was, in Coleman's view, a joke. TWC was emphatically not a joke. The cable television business was booming back in the early 1980s; CNN, the twenty-four-hour news station, was off to a smashingly successful start; so why not a weather channel? To clinch the deal, Coleman came up with a creative way of getting local forecasts to viewers: the Weather STAR, a communications satellite, processes and codes all of the local (or zone) forecasts that the NWS makes four times a day and beams them out to cable companies nationwide. On average the NWS issues between 800 and 1,100 different zone forecasts each day, so the Weather STAR's microchips really hum (the number of zone forecasts varies depending on the weather —tranquil weather covering a large area means that many zones will have the same forecast). When you watch TWC, these local forecasts pop up on your screen every ten minutes or so as teletext messages accompanied by a sound track of soft-lite music. NWS severe weather warnings appear in the same teletext format (minus the music). A newer version of the Weather STAR has added graphics and animation, so now the local forecast segment also includes radar imagery showing the location and movement of precipitation in your area. A national network broadcasting out of one location with detailed, constantly updated local coverage: TWC had broken into major new territory.

The NWS embraced TWC with open arms, even carrying out a major format change in the zone forecasts so they could be transmitted by the Weather STAR's computer; and, after a certain amount of skepticism and ridicule, television viewers also warmed up to the concept. Initially Americans didn't realize that they *needed* round-the-clock weather reports, and the network reportedly plunged into debt to the tune of $7 million after its first year. There was also a good deal of friction over management and control between John Coleman and Landmark Communications, which put up the money to launch TWC, and Coleman was forced out at the end of the first year of operation. "It took more resources than John and I and others thought it would take to get it off the ground," recalls Joseph D'Aleo, who was involved in the creation of the channel but has since left to take a position with WSI. "We were using a thousand hand-produced graphics every day. No one had ever tried to do anything like this before." But eventually things settled down, audiences grew steadily, and by 1986 TWC was turning a profit. It is now, of course, a resounding success— practically an American institution. Close to 90 percent of all cable systems carry TWC, feeding it into 56 million American households. At any given time, the television sets in nearly 200,000 American homes are tuned to TWC—and "ratings soar," reports the New York *Times,* "whenever Big Weather comes along, like the ferocious Northeast storm in December [of 1992] or Hurricane Andrew." Profits have also been soaring. Though Landmark Communications won't release any numbers, the Atlanta-based TWC is in "very good financial shape" according to Ray Ban, vice president of meteorological operations.

"You're watching the Weather Channel," the relentless tag line reminds you when you tune in. "Weather you can always turn to for accurate and dependable weather forecasts twenty-four hours a day." *Dependable* is the key word in the channel's approach to the atmosphere. The folks at TWC are totally straight about the weather: there's no horsing around on the air, no blinding smiles or oily promises of a spectacular tomorrow, no props or pranks, and only a bare minimum of chitchat between "anchors." Three-quarters of the channel's twenty-four on-camera weathercasters are degreed meteorologists. An additional staff of forty-two meteorologists toils behind the scenes preparing forecasts and working with graphic artists on the 800 to 1,000 different images broadcast each day. In the course of an hour, the weathercasters cycle briskly through the various segments —the overall national view at the top of the hour, the "five-day business planner," the Michelin drivers' report, the ski or beach and boating reports depending on the season, the tropical update, the Florida report, the international report, and of course, at regular intervals, "your local forecast, accurate and dependable from TWC." It's a ceaseless oratorio always humming away on the same set of tunes.

The professional meteorologists and weathercasters I spoke with were unanimous in their praise of TWC. Bob Copeland, recently retired after thirty-six years of weathercasting in Boston, says he used to tune in to TWC for a couple of minutes each morning at 3 A.M. before heading in to the studio to do his own morning weather report on WCVB. "That was my early morning weather briefing. I got to see if my previous day's forecast was on target. These folks have national weather down to a science. I still flick it on whenever I want to get an update. The best information they provide is the live radar from your own area every five minutes. If you have half a brain in your body you can tell if rain is coming your way."

Lots of nonprofessionals use the channel in the same way: a quick couple of minutes before they head out the door in the morning. But a surprising one out of five Weather Channel viewers hangs in there for twenty-seven minutes or more, according to research conducted by the A. C. Nielsen Company. As the New York *Times* pointed out in a recent article on the channel, twenty-seven minutes is "enough time to hear the local weather report four times, as well as to catch the regional forecast, the national radar, the school day report, the business traveler update and the forecast for Milan."

Maybe this depth of devotion is not really so surprising. There is something very lulling, very reassuring about TWC. It makes you feel as if there is nothing going on in the world *but* weather. Television news these days is "mostly murders, followed by mostly cloudy," as one weathercaster put it. TWC eliminates the murders. "Some of our viewers want to get the local forecast as quickly as possible and move on, but there are a lot of people who really enjoy TWC," notes Ray Ban. "They watch because it's nonthreatening and noncontroversial. There is no sex or violence on TWC. It provides people with safe programming." True, weather kills too— but a tornado or blizzard or wildfire started by a lightning strike is an act of God and thus nobody's fault. Weather is not evil. TWC scrubs the slate clean of human depravity, waste, and cruelty. It restores us to a televised Eden before the Fall—to a bright, fresh, occasionally stormy world of schoolchildren and picnics, beach trips and tropical vacations, mountain snow and sunny winter days in Florida. Sometimes the commute to work gets messed up by ice or flooding, but that's about as "real" as things get. Even killer weather seems remote and abstract on TWC, unless, of course, it is bearing down on your zone.

The maps are the best part of all. On TWC the outline of the lower forty-eight states becomes a frame for a work of kinetic art in continuous progress: bands of color swirl and eddy across the continent, chilly blues nudging into warm yellows and hot reds, blots of wet green seeping from the Pacific Northwest eastward into the Rockies, troughs chasing ridges in a ceaseless round of "systems." TWC puts the maps in motion with a vengeance —zap! right before your eyes the next five days flash by in a couple of seconds. The weekend begins on Tuesday and climaxes on Thursday afternoon. That storm brewing in the Gulf of Mexico gets itself "organized," sails up the East Coast, and flies out into the Atlantic before you can blink. Instant or almost instant weather gratification: you like to ski? stay tuned and we'll show you where the flakes are flying; plagued by incessant rain? check out this ridge of high pressure sitting over southern California; too hot down there in Texas? take a look at these morning lows in Minnesota! On TWC all of America becomes your backyard. No matter where you live, you get the thrill of a tornado watch, the tranquillity of a cool clear evening, the enchantment of a Christmas snowstorm—any time of the day or night. "Weather you can always turn to," that tag line promises—*turn to* like an old trusted friend. A shoulder to cry on, a steady head in time of crisis, a voice of reason crying out in the wilderness. All this and your local forecast too. Why watch anything else?

TWC FANATICS MAY BE ON THE RISE, BUT SO FAR JOHN COLEMAN'S prediction that TWC would put local weathercasters out of business has not come to pass. "As long as there is local news there will be a local weather report," insists Tulsa's Gary Shore. "The local weathercaster's reputation is based on accuracy and on a personal relationship with viewers. They feel comfortable with you as they get to know you, they trust you. The human element still enters in."

Allan Eustis agrees: "I think the weatherman will remain an institution because of the entertainment value. You know, the attitude of 'Let's laugh at the weathercaster because he's going to be wrong again.' The humor will persist. Local news is so depressing, so why not have some humor?"

The long-running joke of television weather may be too good to pull off the air, but in the future, chuckling local weathercasters will have to compete with no-nonsense, high-tech, high-speed sources of weather information. Three-dimensional displays that give viewers the sensation of flying around and through the weather are already in development, and on-line weather information is becoming increasingly available. If you've got a modem on your home computer you don't even need TWC: you can plug right into the NWS maps and forecasts or get them funneled through one of the private weather companies. "I can foresee virtual-reality forecasts," says Eustis. "By the year 2000 viewers will be able to get weather whenever and wherever they want it." Anticipating these trends, TWC is already exploring ways to ramp on to the information superhighway with "interactive" weather forecasts and video products that show weather in real time as it is unfolding.

Increasingly fast, increasingly localized, increasingly specific weathercasts are the wave of the future. And, of course, increasingly accurate. To quote TWC's Ray Ban, "We'll be able to predict that thunderstorms will move into the north part of your county between 1 and 3 P.M. And we'll be right 99 percent of the time." Wow. But will near-perfect accuracy mean the death of weathercasting comedy? Will we stop laughing if the weatherman is almost always right? Somehow I doubt it. The humor of television weather lies not in its inaccuracy but in its anomaly—in the absurdity of trying to squeeze something as vast and strange and slippery as weather into the square confines of a video screen. It was the genius of the television weatherman to recognize the comedy of the situation and run with it. Great material popping literally out of thin air and a fresh new punch line every night. My guess is that as long as there is television, this cosmic joke will keep on playing from coast to coast.

8.

WHAT'S HAPPENING TO OUR WEATHER?

REMEMBER THE SUMMER OF 1988? TEMPERATURES IN THE EASTern half of the nation hit the 90s by May and pretty much stayed there all summer, except when they climbed into the 100s. As temperature records were broken throughout the Midwest, crops shriveled in the fields. It was not only hot but dry too—bone dry for months on end with no relief. By August barges were stranded in the shrunken, sluggish Mississippi and river commerce ceased. Fires burned out of control in Yellowstone National Park, consuming half of its territory. Forty percent of the counties in the United States were declared disaster areas due to drought. "Electric power faltered, vast forests went up in flames, river navigation was throttled, crops failed," summed up the New York *Times*. By the end of the summer scientists reported that worldwide temperatures had reached their highest levels in 130 years, which was as far back as the records went.

But statistics alone do not capture the eerie sense of doom that settled over the nation like ash that summer. There was something vengeful and apocalyptic about the weather. The brassy sky, sucked dry of moisture, looked cruel and menacing. Cities were beastly outside the air-conditioned enclosures, but even the countryside afforded no escape in much of the nation. Woods, farms, and fields felt like parking lots—parched, smudged, exhausted, pulsing with heat. "What's going on?" people asked each other in alarm, taking for granted the fact that *something* horrible was "going on" and that we as a species were somehow responsible. For a lot of us, the summer of 1988 marked a loss of meteorological innocence and the beginning of a new era of weather consciousness—call it the greenhouse era. "It is all but impossible, we gather, to show that the greenhouse effect 'caused' this summer's weather," *The New Yorker* magazine noted cautiously on August 29, as the summer of 1988 shuddered to a close. "But the weather has obviously caused believers in the greenhouse effect to come out of the closet and declare themselves." The unsigned "Talk of the Town" piece concluded on an ominous note: "Weather talk is the classic example of small talk. A few years from now, we may look back on the summer of 1988 as the moment when it started becoming big talk."

Those few years have passed and weather talk has indeed become big talk, shooting up to enormous proportions with every fresh weather disaster and every record-breaking year of heat. The greenhouse effect has come "out of the closet" with a vengeance since the long hot summer of 1988, and nowadays even elementary school children can rattle off the theory of global warming: that certain gases—especially carbon dioxide—form a kind of glass roof in the upper atmosphere, trapping the earth's heat and radiating some of it back to the ground, and that the greatly accelerated emission of these gases due to industrialization may result in drastically higher temperatures worldwide. "This is a problem the likes of which this planet has never seen before," states Jerry Mahlman, the director of the NOAA's Geophysical Fluid Dynamics Laboratory (GFDL) in Princeton, New Jersey, one of the major centers of climate research.

Concurrent with the record-breaking heat of the 1980s and early 1990s, there has been an epidemic of ferocious weather in just about every part of the country: some of the worst hurricanes this country has ever known—Andrew, Hugo, Gilbert, and Gloria in 1992, 1989, 1988, and 1985, respectively; the midwestern floods of the summer of 1993, in which 10 to 15 million acres were submerged, 48 people died, and over $12 billion in damage took place; the Superstorm of March 13–14, 1993, and the severe winter that hit the Northeast the following year; heavy rains flooding Georgia and Florida during the summer and fall of 1994 while extreme drought and high heat turned the West into a tinderbox, with 3.5 million acres blazing out of control in Washington, Idaho, Colorado, and Montana—an area slightly more than half the size of Massachusetts; much of California declared a disaster area that winter when a series of Pacific storms dumped several feet of rain on parts of the state in just a few weeks. Are these deadly weather swings connected with the onset of the greenhouse effect? The media has conveyed this impression, though Mahlman and other climate theorists emphatically dismiss it as a myth. Nonetheless, many people just can't shake the notion that the weather is turning strange. Violent, unstable, weirdly careening from one extreme to another, and above all hot.

Whether it's happening already, about to happen, or a serious miscalculation on the part of the climatology community, the greenhouse effect has already had an important effect on *us*. Our weather consciousness has altered: the way we perceive, report, and talk about the atmosphere has shifted. As prominent greenhouse theorist Stephen H. Schneider writes of the media coverage of the summer of 1988 in his book *Global Warming: Are We Entering the Greenhouse Century?*: "The daily newspaper, newsweekly, and broadcast stories on the heat, drought, forest fires, and super hurricane Gilbert were overwhelming factors in bringing climate-change issues to the public consciousness."[1] Consider this lead sentence in a piece in the New York *Times* linking that summer's brutally hot weather to greenhouse effect: "The memorably uncomfortable summer of 1988 has left many Americans with a suspicion that nature is at last getting even for mankind's wanton pollution of the atmosphere." Suspicion. Getting even. Wanton pollution. This is the language of guilt and retribution. The *Times* piece reflects a new and pervasive conviction that the strange weather of the 1980s and 1990s is *our fault.* Our waste, our industrialization, our heedless squandering and tainting of the earth's resources have at last caught up with us. We got away with it for a century or more, but we are finally being punished for our sins against the planet. It's the old Puritan theological meteorology, but with a modern, environmentalist, popular science twist. We have encroached upon the upper atmosphere, the last frontier: we have fouled the Olympian realm of chaos and toppled the final stronghold of the balance of nature. And for this we must pay. Bill McKibben speaks for many when he writes in *The End of Nature,* a doomsday treatise about the devastating effects of global warming on our personal and spiritual lives: "We have changed the atmosphere, and that will change the weather. The temperature and rainfall are no longer to be entirely the work of some separate, uncivilizable force, but instead in part a product of our habits, our economies, our ways of life. . . . The world outdoors will mean much the same thing as the world indoors, the hill the same thing as the house." Shades of Increase and Cotton Mather. Trembling awe and bewilderment and guilt and a feeling of doom in the face of violent atmospheric swings crop up repeatedly in our history. America's weather today is a compound of poetry and science, fantasy and expectation, conviction and delusion just as it has always been. What's different, perhaps, is our sense of extremity—the idea that time is running out, that we are approaching the point of meteorological no return.

The new weather consciousness is a product of postindustrial guilt, press hysteria, nature worship, and evangelical anxiety about the approach of the apocalypse. It yokes together scientists and millenarian preachers, radio pundits and backyard gardeners: all interpret the intense weather of the past decade as a *sign,* though they can't agree on a sign of *what.* As the obsession with weather mounts, weather stories have moved to the front page of newspapers and magazines, and have jumped from the two-and-a-half minute slot on the evening news to become the subject of full-scale documentaries. Weather is not only big news, it is politics as well: "the warming" has a full slate of policy implications for industrial development, energy use, and farming techniques, and one's opinion about the validity of greenhouse theories places one squarely in one political camp or the other.

The media hype, for all its in-your-face shrillness, has had the odd side effect of pushing weather closer to the center of our lives, at least some of our lives. Now that we seem to be on the verge of wrecking the weather, we are enshrining it as a kind of wilderness—an ineffable, intangible wilderness of space and time, as magnificent as our mountains and deserts and shorelines, and for all its deadly power, just as fragile. We are coming to see weather as one of our most vital connections to the glory and the mystery of nature. We are recognizing more and more that not only does the atmosphere surround us but it also envelops and infuses and interpenetrates our very being. "The air, in fact, is not 'something apart,' " writes John Firor in *The Changing Atmosphere;* "it is the creation of ourselves and all other living things. . . . All living things combine to establish the composition of the atmosphere; changes in the atmosphere affect life everywhere."[2] It is no coincidence that a new breed of weather fanatics—storm chasers, tornado watchers, blizzard enthusiasts—has arisen at the same time as the theories of global climate change. There's a feeling among these folks of "let's get in on the action before it's too late." They want a close encounter with the wildest brain waves of Mother Nature. Weather on the edge. A peak experience attainable otherwise only on remotest mountaintops or fathomless oceanic depths.

IN THE PUBLIC IMAGINATION, THE GREENHOUSE EFFECT LOOMS as one of the evil specters of the late twentieth century—a member of the same hellish family as ozone depletion (with which it is often confused), AIDS, air and water pollution, and overpopulation. What many don't realize is that the greenhouse effect is in fact a blessing: the naturally occurring (as opposed to man-made) greenhouse effect actually makes life as we know it possible on earth. If there were no greenhouse gases in the upper atmosphere catching some of the infrared energy radiating off the ground and deflecting it back down again, the earth's average temperature would plummet from its current level of 59° F to about 0° F. The complete absence of greenhouse gases would be like trying to get a furnace to heat a house without a roof. So it's not the greenhouse effect per se that is the villain—but *too much* greenhouse effect resulting from the *wrong type* of greenhouse gases: the natural greenhouse effect arises largely from water vapor; the "unnatural," bad, man-made greenhouse effect that may be causing global warming comes from the massive and accelerating increase of carbon dioxide, methane, nitrous oxide, and chlorofluorocarbons (CFCs).

In a sense, the greenhouse effect is a by-product of air pollution, a diseased stepchild of the Industrial Revolution. The greenhouse theory itself was first hatched at the height of the Industrial Revolution, when the skies of Western Europe were heavy with the fumes of coal, wood, and charcoal. In 1896 a Swedish chemist named Svante Arrhenius published a paper postulating that the buildup of carbon dioxide in the atmosphere from all the coal smoke might raise the world's temperature. Arrhenius even came up with an estimate of how high the world's temperature would climb: between 1.5° and 5.5° C if carbon dioxide levels doubled. Scientists have since added refinements and devised complex models to recalculate the correlation between temperature rise and carbon dioxide release, but the underlying concept of the greenhouse effect has remained the same during the intervening century.

What *has* changed is the attention and respect accorded the greenhouse theory. Arrhenius went on to win a Nobel Prize in chemistry in 1903, but his global warming paper went largely unheeded. The increase in carbon dioxide he described was impossible to measure precisely with the technology of the day; and to the extent that people in Europe and America noticed any rise in temperature, they welcomed it—after all, the Little Ice Age had only recently relaxed its grip. There were little ripples of interest in global warming during the 1920s and 1930s (including the 1938 work of a British engineer named G. D. Callendar, who fervently embraced greenhouse warming as a boon to agriculture and a way to block the "return of the deadly glaciers"), but by the mid-1940s the subject had lost its urgency due to a change in the weather. After the hot, dry dust bowl period of the mid-1930s and a few lingering years of warmth in the early 1940s, global mean temperatures began to fall, especially in the Northern Hemisphere. Cooler temperatures persisted through much of the 1950s and returned in the mid-1960s. By the late 1970s, record-breaking cold winters were making headlines in the United States: the coldest winter on record in the Midwest and freezing temperatures in Florida in 1976–77; severe winter storms crippling East Coast cities in 1977–78; new records for cold set in the Southeast. According to some outspoken scientists, a new ice age had begun—or rather the balmy interglacial period was coming to an end. Others attributed the global cooling to an increase of dust particles and smoke in the atmosphere due to rampant industrialization and the spread of agriculture to previously undeveloped parts of the world: the man-made pollutants were thought to be blocking enough of the sun's light to lower temperatures, just as the huge volume of volcanic dust that spewed out from the eruption of Mount Tambora was believed to have precipitated the chilly summer of 1816.

Things got so dire (or at least were *perceived* as so dire) that the CIA jumped in, commissioning two reports in the mid-1970s exploring the "threat" that America faced when global cooling caused "grave shortages of food" in the less developed countries and led to "massive migration." Popular science writers were not far behind, and the decade closed with a slew of books describing global cooling as an established fact and painting grisly doomsday scenarios of the frigid years to come. Books with titles like *The Cooling, Ice or Fire, What's Wrong with Our Weather* warned of worldwide famine, dangerous social upheaval, mass migrations, a crisis in agriculture (in *Ice or Fire,* D. S. Halacy, Jr., proposes growing mushrooms on a vast scale as one possible solution to the problem of cold-climate farming).

Research on the greenhouse theory did not cease in the chilly late 1970s, even though not many people outside of climatology circles were paying much attention to it. In fact, as one climatologist puts it, the really serious scientists had their money on global warming all along. But as long as cold winters hung in over the Northern Hemisphere, greenhouse theorists pretty much kept their heads down, quietly amassing data, running increasingly complex computer models, publishing papers, attending conferences. Even though the most immediate and obvious climatic evidence did not yet support their basic assumption, the warming advocates were nonetheless convinced that the unchecked rise of atmospheric carbon dioxide was too startling to ignore. In 1958, a chemist named Charles David Keeling began to take continuous readings of the absolutely clean air that circulated at 11,050 feet up the flanks of an extinct Hawaiian volcano named Mauna Loa. Keeling, who is still at it, found that carbon dioxide levels were rising steadily over the years—from 315 parts per million in 1958 to 355 parts per million in 1990 to 359 today (autumn 1994). This is the so-called Keeling Curve that one science writer believes "may prove to be a central symbol for the Anthrocene Age, our legacy on earth."

By the start of the 1980s, as global temperatures inched upward, the greenhouse theorists began to seek a wider public forum for their ideas. In 1981, a team of scientists from Harvard University and the Goddard Institute for Space Studies headed by W. C. Wang and James Hansen published the prediction that "the combined warming of carbon dioxide and trace gases" would begin to be detectable *that decade,* with global mean temperatures rising above the heat records set in the late 1930s. This was going way out on a limb at a time when the bitter winters of the late 1970s were still fresh in people's memories and the paperback editions of the cooling titles still gleamed on bookstore shelves. But as all the world now knows, the Wang-Hansen long-range forecast proved to be eerily accurate. The 1980s turned out to be the hottest decade on record, and the 1990s may well be even hotter. For those keeping score, 1995 holds the all-time world heat record, followed by 1990, 1991, 1988, 1981, 1987, 1983, 1980, and 1989.[3] Is this "merely" a temporary shift in the weather —or is it hard evidence of climate change? That question has yet to be resolved, but the 1980s definitely marked a major change in the climate of science. As cultural critic Andrew Ross notes perceptively in his book *Strange Weather,* "In the space of little over a decade, the theory of global cooling, once a dominant thesis among environmental scientists, has been relegated to the margins of legitimacy, espoused today only by crackpots and conspiracy theorists who are ridiculed for their views, as if they still believed the earth is flat."[4]

So abrupt was this turnaround that historians have even pinpointed the exact day on which public opinion shifted from global cooling to global warming: it was June 23, 1988, when James Hansen appeared in front of the Senate Energy Committee and intoned that it was "time to stop waffling so much and say that the evidence is pretty strong that the greenhouse effect is here." Hansen went so far as to throw out a statistic for the media to sink their teeth into: he was "99 percent" sure that the warming of the 1980s was not some fluke of the jet stream but was in fact the result of the buildup of greenhouse gases. "Immediately that '99%' was everywhere," reports Schneider in his book *Global Warming*. "Journalists loved it. Environmentalists were ecstatic. Many meteorologists were upset." Upset or not, global warming became a household phrase overnight, and Hansen was anointed high priest of the greenhouse cult.

The warming story played in the media on a track already well grooved by the cooling scare of the previous decade (though the pace of coverage was a bit quicker now that the computer age was in full swing): first came the front-page headlines and feature stories in science sections, then the "in-depth" interviews and profiles, next the television documentaries, and finally, after a lag of a year or two, the books. Stephen H. Schneider led the pack with his *Global Warming*, published in 1989. Schneider, a scientist with a flair for dramatic prose, opens with a "plausible scenario" of what life in America would look like during the "greenhouse century"—smoke from immense fires in Saskatchewan darkens the daytime sky in the Midwest, 600 people die of heat stress in New York City after twelve days of temperatures over 95°, a near-record hurricane fueled by "the anomalously warm Atlantic" causes massive destruction along the East Coast, prolonged heat and drought lead to the collapse of farming in the corn belt, rapid "dieback" due to mild winters decimates northern forests, chronic water shortages cripple southern California. "Although one can't prove beyond doubt that the greenhouse century has already begun," writes Schneider, "I strongly suspect that by the year 2000 increasing numbers of people will point to the 1980s as the time the global warming signal emerged from the natural background of climatic noise."[5]

A slew of warming titles followed in 1990, which also happened to be the hottest year on record. In *Dead Heat: The Race Against the Greenhouse Effect,* Michael Oppenheimer and Robert H. Boyle declared on page 1, "Humanity is hurtling toward a precipice. Left unchecked, the emissions of various gases, particularly carbon dioxide from fossil-fuel combustion and deforestation, are likely to alter the Earth's climate so rapidly and so thoroughly as to destroy much of the natural world and turn the world that we call civilization upside down." *Greenhouse Glasnost,* edited by Terrell J. Minger, came anointed with an introduction by Robert Redford: "We've always thought of climate as an act of God. It requires an enormous shift in the way we think of the world and our place in it to understand that we have already moved into an era in which we are actually responsible for managing climatic parameters." In *The Greenhouse Trap: What We're Doing to the Atmosphere and How We Can Slow Global Warming,* Francesca Lyman warned, "If the world heats up as much as some scientists now fear, the change will alter virtually every facet of contemporary life, playing havoc with farming, forestry, fishing, transportation, water supplies, and energy . . . There's no aspirin for our planet's fever. The only way to control the greenhouse problem is to cut fossil fuel emissions."

And so on. It's possible this outpouring of titles says as much about book publishing practices as it does about the state of the world's climate —but in any case, after 1990 you'd have to be severely out of the loop not to know about the greenhouse effect. The concept was firmly lodged in the public's imagination and there it remains.

THE INTERVENING YEARS HAVE BROUGHT AN AVALANCHE OF fresh new greenhouse books, along with thousands of articles in newspapers and magazines. When I ran the words "greenhouse effect" through the newspaper and magazine data base of my local library, I came up with 1,100 articles on the subject published since 1989; "global warming" turned up over 1,800 articles. For better or for worse, the greenhouse effect is now part of our lives and part of our weather: we can't hear phrases like "record-breaking heat" or "rising sea levels" or "carbon dioxide emission" without thinking of the greenhouse effect. It is the ascendant star in our skies. The very term now emits an aura of social, cultural, economic, and political implications. Especially political. Climate is politics these days, just as astronomy was religion back in the Renaissance when Galileo peered at the heavens through a telescope. If you know an individual's political leanings, you can pretty well predict his or her views on the global warming threat. "Liberals" embrace it because controlling or reducing greenhouse gas emissions is good for the environment: it's ecologically sound. Conservatives hate it because those same controls are bad for business: they would cost governments and corporations money, raise taxes, hamper industrial productivity, and generally thwart economic growth.

About the only thing the two sides agree on is that the world has indeed gotten measurably warmer in the last hundred years—the number most commonly given is a $1°$ F $(0.5°$ C) rise in mean global temperature since 1890.[6] There is also a broad consensus that greenhouse gases have been building up rapidly in the atmosphere since the start of the Industrial Revolution: carbon dioxide levels have increased about 22 percent in the past century alone and methane levels have doubled; and further, if current rates of emissions continue, in 50 to 100 years our atmosphere will contain twice as much carbon dioxide as it did before the Industrial Revolution. Rising temperatures; increasing levels of greenhouse gases: these are the established "facts." But interpreting what these facts mean, how and if they correlate, and what if anything this correlation portends for the future, have become very controversial topics indeed.

And so the climate change debate rages on, with the media avidly fanning the flames. We now seem to have entered the backing-and-filling phase of global warming media coverage. After a run of warm years during which the greenhouse effect was depicted as a real and present danger, editors and producers became skeptical. The story grew stale. In the early 1990s global temperatures dipped downward, perhaps as a result of the eruption of Mount Pinatubo in 1991. The greenhouse effect seemed to be washed up. And it was boring. Everybody had gotten sick of those same old shots of snarled traffic spewing exhaust into the air while a deadly orange sun beat down through a smoggy sky. And so the Backlash was born. In its July 1993 issue, the conservative magazine *Commentary* ran an influential piece called "Greenhouse Anxiety" by Jeffrey Salmon, the executive director of the George C. Marshall Institute in Washington, D.C. The gist of Salmon's argument was that the computer models used to generate the various predictions of global warming are too crude to digest "the myriad factors that influence climate change." Global warming was an "unlikely candidate for the apocalypse," Salmon announced, and he dismissed the entire issue as a "scare" hatched by "the scientific elite" for personal advancement, media attention, and political manipulation. He concluded (with dubious logic) that since we can't trust the weatherman to get the weekend forecast right, we would be fools to put any faith in greenhouse climatologists: "Everyone knows forecasting the weather more than a day ahead is a hit-or-miss proposition. Why then should we take seriously weather predictions years into the future?" The entire field of atmospheric science blown off in two sentences!

Several other anti-greenhouse theory pieces came out around the same time as Salmon's article, and that September the New York *Times* felt that the dissenting viewpoint had attained the critical mass to qualify as a trend. "Scientists Confront Renewed Backlash on Global Warming" was the headline of a major story by William K. Stevens that ran in the Science section on September 14, 1993. The piece reported that "in a drumroll of criticism" climatological conservatives and their supporters in industry have attacked the greenhouse theory as "a 'flash in the pan,' 'hysteria,' 'scare talk' and a ploy by socialists to justify controls on the economy." Stevens concluded cautiously, "Global warming is not a cut and dried issue, and scientific experts are still debating most of its aspects."

But recently the winds of opinion have shifted once more. Greenhouse theorists predicted that once the volcanic ash spewed out by Pinatubo had settled, the warming trend would continue—and 1994 proved them right. And so global warming is back in the news again—not on front pages, but featured prominently in science sections. It's become one of those familiar seasonal stories, like testimonials to Santa Claus or the bracing autumn aroma of burning leaves, hauled out whenever the weather is suitable.

"DEBATE" IS A WORD THAT COMES UP IN NEARLY EVERY NEWS
story about the greenhouse effect, but curiously enough, climatologists
smile when you ask them which "side" they're on. "The media have set
this issue up as a high-amplitude, high-voltage debate," Mahlman of the
GFDL says, "but the reality is quite different in the climatology community. We talk about what we're unsure about. We know the greenhouse
effect is there—the question is how you quantify it. This is not some hug-
a-tree liberal jive. It's real." Or as Stephen Schneider wrote recently,
"There are vastly greater disagreements over what to do about the prospect
of global warming (i.e., a political value issue) than over the probability
(i.e., a scientific debate) that unprecedented climate change is being built
into the 21st century climate." In Schneider's view the media coverage of
the issue "frequently polarizes scientists into the 'end of the world' or
'nothing bad will happen' factions, even though the scientific community
at large views these as the least probable outcomes. This false dichotomy is
followed by policy gridlock."[7]

In other words, the global warming picture looks completely different depending on where it's being projected: on television screens, in the
marble chambers of government, or in the gleaming computer labs of
university and government research facilities where climate science happens. In the atmosphere of the lab, rhetoric and oratory are turned down
low. Numbers prevail. Variables are isolated and analyzed. Uncertainty is
not a source of confusion but an element to quantify and factor in. Climate
science these days is largely a science of the future: What will happen
when? How much will happen? What will make it happen faster? How
can we stop it from happening? These are huge questions, and huge quantities of money and manpower have been allocated to answering them. The
generous funding has helped make climatology a healthy, even thriving
science, with several strongly promising avenues of research. As David
Schimel, a scientist on staff at the National Center for Atmospheric Research (NCAR), puts it, "Underneath all the hoopla, we are forging ahead.
Good work is being done. We are reducing uncertainties, or at least gain-

ing a better understanding of them." If we want to understand what's happening to our weather these days, we need to dim out the media debate for a while and take a look at the future through the eyes of climatologists.

The good work being done in climatology these days—in fact, just about the only important work—is being done with computer-generated numerical models. A climate model works along pretty much the same lines as a weather forecasting model (see Chapter 6) only on a far vaster and more complicated scale. This stands to reason since a climate model, if it's going to tell us anything worthwhile, must encompass changes not only in the atmosphere but also in the ocean, the polar ice caps, and the biosphere (living organisms and the environment that sustains them) and must span not hours and days but decades and centuries. The very concept of modeling climate is still relatively new, dating back only to the early 1960s when the first climate models were run at Princeton's Geophysical Fluid Dynamics Laboratory; other early climate modes were designed at NCAR, NASA's Goddard Institute for Space Studies, and the British Meteorological Office. In the past decade or so, climate models have proliferated, though all of them are essentially variations on these three or four initial themes. In a relatively short time, modeling has revolutionized the nature of climatology—indeed the models have altered the nature of climate itself. As Thomas Levenson writes in his book *Ice Time,* "Climate, the actual climate out there in nature, has been transformed by the science that studies it. It has become since 1945 an integrated system, a weave of processes whose workings spread through time, over billions of years, to produce the world we live in. The science expanded the concept of climate to make it reach out through nature, from the sun to the bottom of the sea and back again. The concept of climate has also transformed the science and the scientist."[8]

The two hottest areas of modeling now are clouds and the ocean. The ocean is vital to calculating climate change because it has essentially all the heat capacity of the system. Jerry Mahlman explains: "Take a bathtub and fill it to within one-tenth of an inch of the top. That top layer is the atmosphere. Heating the ocean is like trying to heat that bathtub by shining a night light at it. There is easily 1000 years of heat capacity in the ocean." What impact will this heat-absorbing capacity have on global warming? How deeply into the ocean will atmospheric warming penetrate —and how quickly? How will greenhouse warming affect major oceanic currents like the Gulf Stream? How much carbon dioxide will the ocean be able to absorb? What impact will changes in the ocean have on atmospheric temperatures and circulation patterns? These are some of the questions that climatologists are trying to resolve through computer models.

Dr. Syukuro Manabe, with the GFDL, is the acknowledged pioneer in this field and he continues to dominate it. Manabe and Kirk Bryan, an ocean modeler, ran the first coupled ocean-atmosphere model way back in the mid-1960s, years before anyone else even thought this could be feasible. It was a triumph of a still-infant science, but a rather crude and primitive triumph—an "inspired approximation," as one of Manabe's colleagues puts it. Only in recent years have climatologists been able to couple fully dynamic three-dimensional oceanic and atmospheric models, but the difficulties are still fairly daunting. As one researcher explained to me, "We know the oceans; we know the atmosphere. But when we try to couple them, the model goes into nonlogical stages. The first test of a model is to see if it can reproduce the current climate, and we haven't been able to do this yet with the coupled ocean-atmosphere model. This remains a huge problem." Climatologists are convinced that their results will improve dramatically once they can get their hands on a new generation of supercomputers. In any case, this is certain to be a focal point of research for years to come. As one climatologist writes, "Understanding the coupling of these systems [ocean and atmosphere] is crucial to projecting the climate of the 21st century."[9]

ENSO—the El Niño Southern Oscillation—throws in another oceanic wild card. An ENSO event is a periodic (every four to seven years) alteration of the ocean-atmosphere interaction in the equatorial Pacific: the strong trade winds, which ordinarily force warm Pacific waters to the west, die down and the surface waters grow warmer, which in turn causes a further weakening in the trade winds. As this "positive feedback loop" intensifies, winter weather patterns undergo a significant shift over much of the world: here in the United States, an ENSO event makes the Southeast wetter and cooler than average, the Northeast and Pacific Northwest become warmer, and California either wetter or drier depending on the position of the jet stream. The most intense ENSO event of the century struck during the winter of 1982–83, with five hurricanes hitting French Polynesia, floods in Louisiana and Florida, and serious drought in Hawaii, Mexico, and Australia. Michael J. McPhaden, who studies ENSO at the NOAA/Pacific Marine Environmental Laboratory in Seattle, notes that these events occur naturally—they are not the result of man-made pollution or heavy submarine traffic. ENSO events just "happen," like the weather, and they have been happening for a long time. We know from records kept by the Spanish conquistadors that they observed the effects of an ENSO event in 1532–33, and the cycle has probably been going on for far longer than that. But there is some evidence that ENSO events are becoming more frequent and more intense, and it's possible that greenhouse warming is to blame. So far no one has been able to explain the unprecedented ENSO of 1991–93, which seemed to be fading but then resurged (causing the massive flooding in the Midwest during the summer of 1993) and which still seems to be lingering on in strange ways (it may be the culprit responsible for the heavy rains that flooded much of California during the winter of 1994–95). McPhaden says cautiously that there is "some indication" that ENSOs will intensify with greenhouse warming, but that "anything is possible." James Hansen notes that the linkage of ENSO events and global warming is "speculative at present," but he adds, "I would guess that the frequency of El Niños may increase, at any rate low-latitude ocean temperatures will increase." Meanwhile, research continues.

Modeling clouds presents a different kind of challenge—in fact clouds are the "single largest source of uncertainty" in current climate modeling, according to Jerry Mahlman. Mahlman notes that the critical question about the ocean relates to the *rate* of warming, whereas the key question in cloud research concerns the *amount* of warming. Will clouds offset or enhance global warming? Climatologists are still unsure. Will clouds have a different impact on warming in different geographical regions? We don't know that one either—but we're working on both. As

Mahlman puts it, "We don't model clouds well, but they're too important to ignore." There is some evidence that the overall cloud cover of the earth is increasing, perhaps as a result of global warming. Richard Lindzen of MIT, an outspoken critic of warming theories, believes there is "compelling" evidence that clouds and water vapor will actually reduce greenhouse warming. Others suggest that the increase of cloudiness is flattening the differences between daytime highs and nighttime lows. Patrick J. Michaels, another greenhouse theory attacker, speculates that the night warming caused by this increase of cloudiness might actually be a boon to agriculture by prolonging growing seasons. This possible benefit leads Michaels to conclude emphatically that "the popular vision of climate apocalypse is wrong."[10] On the other hand, the prospect of living under increasingly cloudy skies is not too appealing either.

Polar ice introduces still other ambiguous factors into the climate change equation. If polar ice melts as temperatures rise, more water vapor will enter the atmosphere, possibly increasing not only cloud cover but also precipitation, including snow in the colder regions; but more snowfall might then replenish or even expand polar ice, thus counteracting the impact of the warming. It's possible that clouds and polar ice could work together to slow the warming or even reverse the trend. Computers are crunching through models incorporating these new variables, but no clear answers have emerged yet.

Polar ice is also a critical source of data in an entirely different line of climatological research. Analysis of ice cores drilled in Greenland suggests that there were considerable and rapid temperature swings during the Ice Age—so rapid that in a decade or less the world grew markedly warmer or colder, which is exactly what happened in the 1970s and 1980s. A recent article in *Scientific American* reports that this finding has prompted some scientists to speculate that "global temperatures might take another violent swing during the current warm spell." It's also possible that this warm spell is something of an illusion—that in fact the climate is *trying* to get colder but human-generated carbon dioxide is masking the cooling. Again, major questions loom, but firm answers remain elusive. As the author of the *Scientific American* article writes, this branch of climatology is "progressing at breakneck speed," but it remains to be seen where the progress is leading.[11]

Climate modelers also have to figure out a way to account for sulfate aerosols—tiny particles spewed into the atmosphere by factory chimneys. Though sulfates are a type of pollution, and thus bad for the environment, they also serve to lower temperatures by reflecting back solar radiation, and thus they may be helpful in counteracting the greenhouse effect. Exactly how much greenhouse warming sulfates will offset and how to balance this benefit against the harmful effects of increased air pollution are questions that climatologists are just beginning to explore. Again, bigger, better, and more sophisticated models are needed to achieve reliable answers.

Finally, on the fringes of science that border on science fiction are the schemes to combat or counteract the greenhouse effect. A few of the more ingenious plans in circulation now include putting huge aluminum foil mirrors into orbit to shield the earth from some of the sun's radiation; diffusing vast quantities of sulfur dioxide into the atmosphere as a way of increasing its reflectiveness and thus bouncing some of the sun's heat energy back into space; fertilizing the oceans so that their increased volume of plant life would absorb more carbon dioxide. I thought these notions sounded pretty bizarre, but when I asked several scientists about them, the major objection they had concerned financing. "Not economically feasible," Hansen told me, "and they don't make sense now given our primitive understanding of the system. But of course we already are 'engineering' the climate, so eventually some counter-engineering may make sense. For the time being we should try to minimize our inadvertent engineering and understand the climate system better."[12]

Climate engineering and counter-engineering sound suspiciously like a couple of new wrinkles on the old pseudoscience of weather modification —rainmaking—which has been kicking around America since before the Europeans arrived. Over the centuries we've tried everything from rain dances to human sacrifice to prayer vigils to seeding clouds with silver iodide crystals, but so far no one has conclusively proved that any of it really *works*. The weather is still the weather—chaotic, unfathomable, infinitely complicated, and ultimately ungovernable, at least by us.

If we can't change the weather, how, one wonders, will we be able to change the climate—or rather, change it back? And change it back to what? What is the normal state of our climate? What is the ideal weather? And who gets to decide? It is my contention that our weather and climate have been strange since the beginning of our history. Our perceptions have always been skewed by expectation, our memories distorted by self-interest. The prime quality of our weather is what it isn't—how it differs from what "true" weather is supposed to be: colder than England, wetter than France, drier in the West than in the East. Nowadays it's just too darned hot just about everywhere—or at least a lot hotter than it used to be in the "old days," when we had real snowstorms every winter and ice-skating on frozen ponds. I'm not saying the greenhouse theory is a kind of collective fantasy, some sort of national or global delusion or a sign that we've all gone soft and just can't take the heat anymore. But I do think that we aren't making enough allowances for natural (as opposed to human-induced) climatic swings. As Mahlman puts it, "Even if you had all the data, you would still have a fierce intellectual challenge to interpret which part of the warming is natural variation and which comes from adding junk to the atmosphere." We're impatient: we want answers and explanations, not fierce challenges. We're too quick to point a finger, too eager to identify a culprit. One climatologist told me, "A lot of modelers want to rush to the White House with their latest printout. They are looking for instant policy change based on their predictions. But prediction without understanding is indistinguishable from fortune-telling."

If you listen to the climate modelers, you get the feeling that we're on the cusp of some crucial homestretch phase of weather history. The next five years, they claim, should tell whether the models are on target. If it doesn't get warmer during the 1990s, writes Stephen Schneider, then we'll know from "direct evidence that the effect predicted by current models is off base." But what if it *does* keep getting warmer? What does this mean about the longer-term future? An accelerating spiral toward climatic apocalypse? Are we heading for "a possible tearing apart of the structures of communities of plants, insects, and animals . . . at rates that exceed clear historic or geologic metaphors," as one scientist puts it? Probably not, most climatologists now feel; the more cautious people in the field see not cataclysm, but rather a gradual but unmistakable erosion of nature as we know it, and thus inevitably of human culture as well. But whatever research paper you read, the same anxiety vibrates beneath the graphs and printouts and comparisons of model runs: We still don't fully understand what, but *something* is happening: we're taking our atmosphere—and thus our species—into new and uncharted territory. As science writer Jonathan Weiner puts it in a recent piece about climate change, "The first dramatic effects of carbon dioxide will probably come where and when we least expect them, because we are playing with something much more complicated than we understand. . . . We are changing the conditions under which we and all the other species on the planet have flourished, and we don't know what will happen next."[13]

My own feeling is that the whole greenhouse issue has been so hyped since 1988 that it has assumed an independent life of its own: the concept has become part of the way we experience and talk about the weather. We don't have storms or heat waves or droughts anymore—we have *mounting evidence* or *ominous signs* or *inescapable indications* or *the onset*. Climate change is in the air, and we don't need the media any longer to point out the global, even cosmic implications of unusual weather patterns. The fact that the major climate story shifted in the course of a decade from global cooling to global warming doesn't bother us at all. The truth is, we've grown accustomed to the idea of climate change—even comfortable with it, for all the possibly imminent horrors of melting or expanding polar caps, super tropical storms or snow blitz, endless summer or eternal winter. Both cooling and warming theories have the ultimate effect of demystifying the weather, of placing it within our grasp. Both theories put humankind, for good or ill, at the center of the atmosphere. Since we caused the climate change in the first place, it stands to reason that we can change it back again. Many of the books on climate change—whether the cooling books of the 1970s or the warming books of the 1990s—conclude with a chapter of "practical steps" that each and every one of us can take right now to slow or reverse the course of catastrophic climate change: plant a tree, take the bus or ride your bike instead of driving your car, recycle, lobby for policy change, work to save the rain forests. Strangely enough, the cooling and the warming books give a lot of the same advice. The underlying message in both cases is, if you can change your life, you can save the world. It's a contemporary environmental version of prayer and good works: "procure a day of humiliation and supplication," advised Cotton Mather in March 1717, so that God would stop sending "such storms and heaps of snow . . . in the approach of Spring"; curb or abandon the use of fossil fuels now, demand Michael Oppenheimer and Robert H. Boyle in their book *Dead Heat,* so that we can limit the warming to just a couple of degrees. Our current obsession with climate change is, in a sense, the culmination of a "trend" that has been building for centuries.

There's a biblical cast to our current thinking about climate: it posits time as a linear progression, and history as a story/sequence with a beginning, a middle, and an end. "Some say the world will end in fire, / Some say in ice," wrote Robert Frost. But why must we choose? And must the world really "end"—or at least turn so beastly awful that it presents an image of that horror? Against this linear model I would hold up the dailiness and seeming randomness of weather. A cycle with infinite variations and endless repetitions that goes nowhere and comes from nowhere. Yesterday's weather—gone without a trace. Tomorrow's weather—a shadow already taking on form and substance and color in the bowels of the supercomputers, but a form that may well assume an entirely different cast when no one is looking. Ten thousand years ago—an ice age. Yesterday—record-breaking heat. Tomorrow—showers? Five years from now—absolute proof of greenhouse warming. Ten thousand years in the future—another ice age?

Linear or cyclical. Meaning or chaos. Our fault. God's fault. Nature's mystery. All of these have been elements of our weather from the beginning. And with all due respect to the scientific community, I'm guessing that they will continue to be. The history of weather is both a history of nature and a history of human desire. A history that is made and erased every day. I believe Dr. Manabe was expressing a similar idea when he told me, "A model is always a simplified version of nature. And one must be humble about whatever happens in nature. Humble and open to challenge."

And with that, I'll close—and leave the weather to you.

NOTES

INTRODUCTION

1. Thomas Levenson, *Ice Time: Climate, Science, and Life on Earth* (New York: Harper & Row, 1989), p. 79.
2. Francesca Lyman, *The Greenhouse Trap: What We're Doing to the Atmosphere and How We Can Slow Global Warming* (Boston: Beacon Press, 1990), p. 26.
3. Jack William, *The USA Today Weather Book* (New York: Vintage Books, 1992), p. 1.
4. James LeMoyne, "In the Storm," *The New Yorker*, October 5, 1992, p. 88.
5. *The Wall Street Journal*, August 8, 1994, p. A5.
6. James Lovelock, *The Ages of Gaia* (New York: Norton, 1988), p. 218.
7. Joseph E. Brown, "Weather or Not," *Travel & Leisure*, October 1992, p. E10.
8. Levenson, p. 79.

1. NATIVE AMERICAN WEATHER

1. Nigel Davies, *Voyagers to the New World* (Albuquerque: University of New Mexico Press, 1979), p. 25.
2. The date of the arrival of human beings in the New World has generated a loud, voluminous, and sometimes acrimonious debate in the past few decades. The dates I have indicated are the estimates of conservative archaeologists and anthropologists; in the 1960s, Louis Leakey blew open the whole subject by claiming that he had evidence for human settlements in

California dating back 100,000 to 200,000 years, and possibly even earlier. Some scholars even claim that the human species *originated* in the New World, migrated to Asia, and then came back. As one "liberal" told me, "You better put these older dates in your book so that you don't look foolish when archaeologists turn up new evidence to back them up." I'm standing by.

3. Philip Kopper, *The Smithsonian Book of North American Indians* (Washington, D.C.: Smithsonian Institution, 1986), p. 33.

4. H. E. Wright, Jr., "Environmental Conditions for Paleoindian Immigration," in Tom D. Dillehay and David J. Meltzer, eds., *The First Americans: Search and Research* (Boca Raton: CRC Press, 1991), p. 119.

5. Warwick M. Bray, Earl H. Swanson, and Ian S. Farrington, *The Making of the Past: The New World* (New York: Elsevier Phaidon, 1975), p. 68. And Jesse D. Jennings, ed., *Ancient Native Americans* (San Francisco: W. H. Freeman, 1979), pp. 29–30, 185.

6. Jesse D. Jennings, "Origins," in *Ancient Native Americans,* p. 11.

7. Jennings, "Origins," p. 37.

8. Brian M. Fagan, *The Great Journey* (New York: Thames & Hudson, 1987), p. 239.

9. Jesse D. Jennings, *Prehistory of North America* (Mountain View, Calif.: Mayfield Publishing, 1989), p. 5.

10. James E. Fitting, "Regional Cultural Development, 300 B.C. to A.D. 1000," in Bruce G. Trigger, ed., *Handbook of North American Indians: The Northeast,* vol. 15 (Washington, D.C.: Smithsonian Institution, 1978), p. 44.

11. Bray et al., p. 80.

12. Kendrick Frazier, *People of Chaco: A Canyon and Its Culture* (New York: Norton, 1986), p. 181.

13. Vincent Scully, *Pueblo: Mountain, Village, Dance* (New York: Viking, 1975), p. 20.

14. Frazier, p. 185.

15. Alfonso Ortiz, Introduction to Alfonso Ortiz, ed., *Handbook of North American Indians: The Southwest,* vol. 9 (Washington, D.C.: Smithsonian Institution, 1979), p. 3.

16. William D. Lipe, "The Southwest," in Jennings, ed., *Ancient Native Americans,* pp. 378–79.

17. Hartley Burr Alexander, *The World's Rim: Great Mysteries of the North American Indians* (Lincoln: University of Nebraska Press, 1953), p. 79.

18. Ortiz, p. 184.

19. Ake Hultkrantz, *Native Religions of North America* (San Francisco: Harper & Row, 1987), p. 108.

20. Louis A. Hieb, "Hopi World View," in Ortiz, p. 577.

21. Hultkrantz, p. 102.

22. Frazier, p. 20.

23. Harold E. Driver, *Indians of North America* (Chicago: University of Chicago Press, 1961), pp. 85–86.

24. As recounted in Gary Lockhart, *The Weather Companion* (New York: John Wiley & Sons, 1988), p. 195.

25. Cited in Kopper, p. 65.

26. Zora Neale Hurston, *Their Eyes Were Watching God* (New York: Harper & Brothers, 1937), p. 153.

2. WEATHER IN THE AGE OF DISCOVERY

1. There have been claims of a number of pre-Norse European voyages to the New World, but historians by and large dismiss them as legends. The Irish proudly insist that the monk St. Brendan sailed to Iceland and possibly on to Labrador and Newfoundland in an open boat—the carraugh made of basketwork and oxhide—in the sixth century A.D. The Welsh also have a contender: Madoc by name, who supposedly set out from Wales in 1170 at the head of a flotilla of ten ships and made land in Alabama. Supposedly the Indians they encountered understood Welsh perfectly—acquiring this notoriously difficult language on the spot. The story of Madoc has persisted into modern times, and the Daughters of the American Revolution have set up a marker at Fort Morgan on Mobile Bay commemorating the Welsh explorer "who landed on the shores of Mobile Bay and left behind, with the Indians, the Welsh language."

2. Quoted in Samuel Eliot Morison, *The European Discovery of America: The Northern Voyages* (New York: Oxford University Press, 1971), p. 46.

3. Reid A. Bryson and Thomas J. Murray, *Climates of Hunger* (Madison: University of Wisconsin Press, 1977), pp. 24–29.

4. Nigel Calder, *The Weather Machine* (New York: Viking, 1974), p. 16.

5. Quoted in Emmanuel Le Roy Ladurie, *Times of Feast, Times of Famine: A History of Climate Since the Year 1000* (New York: Noonday Press, 1971), p. 170.

6. Evelyn Stefansson Nef, "Preface" to Louis B. Wright and Elaine W. Fowler, eds., *West and By North* (New York: Delacorte, 1971).

7. Verrazano, "Report to the King by Giovanni da Verrazano," quoted in Wright and Fowler, p. 75.

8 Quoted in Henry Steele Commager, *The Empire of Reason: How Europe Imagined and America Realized the Enlightenment* (Garden City, N.Y.: Doubleday, 1977), p. 66.

9. Wright and Fowler, p. 104

10. John Bakeless, *America As Seen by Its First Explorers* (New York: Dover, 1961), p. 116.

11. Quoted in David M. Ludlum, *The Weather Factor* (Boston: Houghton Mifflin, 1984), p. 2.

12. Quoted in David M. Ludlum, *Early American Winters 1604–1820* (Boston: American Meteorological Society, 1966), p. 5.

13. Ludlum, *The Weather Factor*, p. 12

14. William Bradford, *Of Plymouth Plantation 1620–1647*, ed. by Samuel Eliot Morison (New York: The Modern Library), p. 62.

15. Bradford, p. 279.

16. Robert Claiborne, *Climate, Man, and History* (New York: Norton, 1970), p. 391.

17. Quoted in Ludlum, *Early American Winters*, p. 11.

18. William Wood, *Wood's New Englands Prospect* (London, 1634), quoted in Ludlum, *Early American Winters*, pp. 13–14.

3. COLONIAL AND REVOLUTIONARY WEATHER

1. Eliphalet Adams, "God Sometimes Answers His People by Terrible things in Righteousness" (New London, Conn.: T. Green, 1735).

2. William Bradford, *Of Plymouth Plantation: 1620–1647*, ed. by Samuel Eliot Morison (New York: The Modern Library), pp. 131–32.

3. Quoted in I. Bernard Cohen, *Benjamin Franklin's Science* (Cambridge, Mass.: Harvard University Press, 1990), p. 244.

4. Perry Miller, *The New England Mind: From Colony to Province* (Boston: Beacon Press, 1953), p. 437.

5. Increase Mather, *Remarkable Providences Illustrative of the Earlier Days of American Colonisation* (London: John Russell Smith, 1856), pp. 51–96.

6. Mather, p. 256.

7. Quoted in David M. Ludlum, *Early American Winters: 1604–1820* (Boston: American Meteorological Society, 1966), p. 43.

8. Miller, p. 443.

9. Mather, p. 88.

10. Letter from Ben Franklin to Peter Collinson, October 1752, quoted in Cohen, p. 68.

11. Andrew Dickson White, *A History of the Warfare of Science with Theology in Christendom,* quoted in James Rodger Fleming, *Meteorology in America, 1800–1870* (Baltimore: Johns Hopkins University Press, 1990), p. 8.

12. Cohen, p. 141.

13. Quoted in Cohen, p. 119.

14. Quoted in Cohen, p. 153.

15. Bradford, p. 303.

16. Quoted in Cohen, p. 141.

17. Ernest Cassara, *The Enlightenment in America* (Boston: Twayne, 1975), p. 49.

18. Cohen, p. 30.

19. Charles A. Miller, *Jefferson and Nature* (Baltimore: Johns Hopkins University Press, 1988), p. 3.

20. Edwin T. Martin, *Thomas Jefferson: Scientist* (New York: Henry Schuman, 1952) p. 131.

21. Miller, *Jefferson and Nature,* p. 41.

22. Alexander McAdie, "A Colonial Weather Service," originally published 1894, reprinted in I. Bernard Cohen, ed., *Thomas Jefferson and the Sciences* (New York: Arno Press, 1980), p. 331.

23. Quoted in Thomas Jefferson, *Notes on the State of Virginia* (Chapel Hill: University of North Carolina Press, 1954), pp. 58–59.

24. See, for example, *Civilization and Climate* by Ellsworth Huntington (New Haven: Yale University Press, 1915).

25. Jefferson, *Notes on the State of Virginia,* pp. 59–60.

26. Jefferson, *Notes on the State of Virginia,* pp. 64–65.

27. Jefferson, *Notes on the State of Virginia,* p. 65.

28. Timothy Dwight, *Travels in New-England and New-York* (London: William Baynes and Son, 1823), pp. 52, 58.

29. Henry Steele Commager, *The Empire of Reason* (Garden City, N.Y.: Doubleday, 1977), pp. 86–87.

30. J. Hector St. John Crèvecoeur, *Letters from an American Farmer* (Gloucester, Mass.: Peter Smith, 1978), p. 53.

31. Quoted in Ludlum, p. 212.

32. C. F. Volney, *A View of the Soil and Climate of the United States of America* (1804), trans. with occasional remarks by C. B. Brown (New York: Hafner, 1968), pp. 262–63.

33. Brown footnote in Volney, p. 107.

34. Jedidiah Morse, *The American Geography* (New York: Arno Press and The New York Times, 1970; originally published 1789), pp. 203, 345, 423–24, 470.

35. Sir Augustus John Foster, *Jeffersonian America: Notes on the United States of America Collected in the Years 1805–6–7 and 11–12* (San Marino, Calif.: Huntington Library, 1954), p. 171.

36. Quoted in Ludlum, p. 247.

37. William Tudor, Jr., "An address delivered to the Phi Beta Kappa Society, at their anniversary meeting at Cambridge," reprinted in *North American Review,* 1816, p. 15.

38. Dwight, p. 48.

39. Cotton Mather, "Winter Meditations: Directions How to Employ the Leisure of the Winter for the Glory of God" (Boston: Benjamin Harris, 1693), unpaged.

40. Jefferson, *Notes on the State of Virginia,* p. 80.

41. Quoted in Volney, p. 214.

42. Volney, p. 222.

43. Andrew Ross, *Strange Weather: Culture, Science, and Technology in the Age of Limits* (London: Verso, 1991), p. 197.

44. Dwight, p. 37.

45. Ludlum, p. 111.

46. Quoted in Ludlum, p. 190.

47. Quoted in Ludlum, pp. 190, 191.

48. Patrick Hughes, *American Weather Stories* (Washington, D.C.: U.S. Department of Commerce, 1976), p. 48.

49. David C. Smith, Harold W. Borns, W. P. Baron, and Anne E. Bridges, "Climatic Stress and Maine Agriculture, 1785–1885," reprinted in T. M. L. Wigley, M. J. Ingram, and G. Farmer, eds., *Climate and History* (Cambridge: Cambridge University Press, 1981), pp. 451–56.

4. WEATHER IN THE WEST

1. Jonathan Raban, "The Next Last Frontier," *Harper's,* August 1993.

2. Bernard De Voto, *Across the Wide Missouri* (Boston: Houghton Mifflin, 1947), p. 2.

3. Marc Reisner, *Cadillac Desert* (New York: Penguin, 1986), pp. 19–20.

4. Maurice S. Sullivan, ed., *The Travels of Jedediah Smith: A Documentary Outline, Including His Journal* (Lincoln: University of Nebraska Press, 1934).

5. Wallace Stegner, "Living Dry," in *Where the Bluebird Sings to the Lemonade Spring* (New York: Penguin, 1992), p. 70.

6. Merlin P. Lawson, *The Climate of the Great American Desert* (Lincoln: University of Nebraska Press, 1974), p. 54.

7. Edwin Bryant, *What I Saw in California* (Lincoln: University of Nebraska Press, 1985), pp. 60–61.

8. Quoted in Lawson, p. 65.

9. Quoted in Bryant, p. 264.

10. Walter Prescott Webb, *The Great Plains* (Boston: Ginn and Co., 1931), p. 157.

11. Wallace Stegner, *Beyond the Hundredth Meridian: John Wesley Powell and the Second Opening of the West* (Boston: Houghton Mifflin, 1954), p. 399.

12. Reisner, p. 42.

13. Hamlin Garland, *A Son of the Middle Border* (New York: Macmillan, 1962), p. 260.

14. Webb, p. 376.

15. Everett Dick, *The Sod-House Frontier* (New York: D. Appleton-Century Co., 1937), p. 223.

16. Dick, *The Sod-House Frontier*, p. 228.

17. W. H. O'Gara, ed., *In All Its Fury: A History of the Blizzard of January 12, 1888* (Lincoln, Neb.: The Jan. 12, 1888, Blizzard Club, 1947), p. 38.

18. John Moses, *Illinois: Historical and Statistical*, quoted in C. Donald Ahrens, *Meteorology Today* (St. Paul: West Publishing Co., 1991), p. 329.

19. Laura Ingalls Wilder, *The Long Winter* (New York: Harper & Brothers, 1940), p. 225.

20. O. E. Rölvaag, *Giants in the Earth* (New York: HarperCollins, 1991), p. 260.

21. Rölvaag, p. 415.

22. John P. Finley, *Tornadoes* (New York: The Insurance Monitor, 1887), p. 7.

23. Finley, pp. 91–92.

24. Quoted in Clark C. Spence, *The Rainmakers* (Lincoln: University of Nebraska Press, 1980), p. 81.

25. John Wesley Powell, *Report on the Lands of the Arid Region of the United States* (2nd ed.; Washington, D.C.: U.S. Government Printing Office, 1879), p. 1.

26. Powell, p. 91.

27. Donald Worster, *Dust Bowl: The Southern Plains in the 1930s* (New York: Oxford University Press, 1979), p. 70.

28. Worster, p. 14.

29. Worster, pp. 13, 24.

30. Worster, p. 49.

31. Thomas Frederick Saarinen, *Perception of the Drought Hazard on the Great Plains* (Chicago: University of Chicago Press, 1966), p. 21.

32. Webb, p. 375.

33. Horace Greeley, *An Overland Journey from New York to San Francisco in the Summer of 1859* (New York: Knopf, 1964), p. 280.

34. Marion Thrasher, M.D., *Long Life in California* (Chicago: M. A. Donogue & Co., 1915), pp. 109, 112.

35. Greeley, pp. 274, 280.

36. Reisner, pp. 344–45.

37. John Muir, *My First Summer in the Sierra* (Boston: Houghton Mifflin, 1916), pp. 235–36.

38. Natalie Angier, "Warming? Tree Rings Say Not Yet," New York *Times*, December 1, 1992, p. C4.

5. FROM PROVERB TO PROBABILITY: WEATHER FORECASTING IN AMERICA

1. New York *Tribune*, August 12, 1895, p. 5.

2. H. H. C. Dunwoody, *Weather Proverbs* (Washington, D.C.: U.S. Government Printing Office, 1883), pp. 5–6.

3. Robin Page, *Weather Forecasting the Country Way* (New York: Summit Books, 1977).

4. George D. Freier, *Weather Proverbs* (Tucson: Fisher Books, 1992), pp. vi, 6.

5. James Rodger Fleming, *Meteorology in America: 1800–1870* (Baltimore: Johns Hopkins University Press, 1990), pp. 143, 145.

6. Quoted in Fleming, p. 23.

7. J. P. Espy, "Theory of rain, hail, snow and the water spout, deduced from the latent caloric vapour and the specific caloric of atmospheric air," as quoted in W. E. Knowles Middleton, *A History of the Theories of Rain* (London: Oldbourne, 1965), p. 157.

8. Quoted in Middleton, p. 160.

9. Cleveland Abbe, *Popular Science Monthly,* quoted in Middleton, p. 160.

10. Quoted in Fleming, p. 79.

11. Quoted in Fleming, p. 78.

12. Address by Professor Thomas Corwin Mendenhall, quoted in Truman Abbe, *Professor Abbe and the Isobars* (New York: Vantage Press, 1955), p. 125.

13. For more detail about the bureaucratic evolution of the Weather Service, see *A Century of Weather Service* by Patrick Hughes (New York: Gordon & Breach, 1970), which traces the bureau from its founding in 1870 to its one hundredth birthday, and *A History of the United States Weather Bureau* by Donald R. Whitnah (Urbana, Ill.: University of Illinois Press, 1961), which goes into more detail but stops in 1961.

14. Cited in Whitnah, p. 163.

15. Whitnah, pp. 86–97.

16. Cited in Whitnah, p. 116.

17. New York *Tribune,* August 12, 1895, p. 5.

18. Whitnah, p. 116.

19. Cited in Bette Roda Anderson, *Weather in the West* (Palo Alto, Ca.: American West Publishing, 1975), p. 172.

20. New York *Tribune,* March 13, 1888, p. 1.

21. New York *Tribune,* March 13, 1888, p. 2.

22. Whitnah, p. 160.

23. Ralph Hardy et al., *The Weather Book* (Boston: Little, Brown, 1982), p. 202.

6. NATIONAL WEATHER NOW

1. Uccellini noted that not even the MRF indicated the severity of the low off the west coast of Florida. Because of this, NWS forecasters underestimated the storm surge that swept up out of the Gulf of Mexico, drowning thirteen people on land and five people whose boats were wrecked (statistics from the *Natural Disaster Survey Report,* pp. 1–2). This was the one major inaccuracy in the NWS forecast of the March blizzard.

2. See, for example, "On the Skill and Utility of NMC's Medium-Range Central Guidance," by M. Steven Tracton in *Weather and Forecasting,* vol. 8 (March 1993), pp. 147–53.

7. THE LONGEST-RUNNING JOKE

1. Jay Rosen, "Don't Need a Weatherman?" *Harper's,* April 1989, p. 34.

2. Robert Henson, *Television Weathercasting* (Jefferson, N.C.: McFarland, 1990), pp. 80–81.

3. Henson, pp. 80–81.

8. WHAT'S HAPPENING TO OUR WEATHER?

1. Stephen H. Schneider, *Global Warming: Are We Entering the Greenhouse Century?* (San Francisco: Sierra Club Books, 1989), p. 191.

2. John Firor, *The Changing Atmosphere: A Global Challenge* (New Haven: Yale University Press, 1990), pp. 1, 4.

3. Andrew Revkin, *Global Warming: Understanding the Forecast* (New York: Abbeville Press, 1992), p. 59, and Stephen H. Schneider, "Degrees of Certainty," *National Geographic Research & Exploration,* Spring 1993, p. 174.

4. Andrew Ross, *Strange Weather: Culture, Science, and Technology in the Age of Limits* (London: Verso, 1991), p. 200.

5. Schneider, *Global Warming,* p. 32.

6. Dissenters question the validity and consistency of instruments used to make readings: some argue that at least part of the apparent warming may in fact be due to the urban "heat island" effect—concrete and brick retain more heat than woods and fields, so as cities have engulfed rural and suburban areas, thermometers left in the same site show steadily rising temperatures.

7. Schneider, "Degrees of Certainty," p. 183.

8. Thomas Levenson, *Ice Time: Climate, Science, and Life on Earth* (New York: Harper & Row, 1989), pp. 114–15.

9. Thomas R. Karl, "Missing Pieces of the Puzzle," *National Geographic Research & Exploration,* Spring 1993, p. 246.

10. Patrick J. Michaels, "Benign Greenhouse," *National Geographic Research & Exploration,* Spring 1993, p. 232.

11. Corey S. Powell, "Cold Confusion," *Scientific American,* March 1994, pp. 22–28.

12. James Hansen, letter to the author, September 30, 1994.

13. Jonathan Weiner, "Winter Forecast: Frigid. But Don't Be Fooled," *The New York Times Magazine,* October 23, 1994, p. 57.

INDEX

ABOUT THE AUTHOR

David Laskin is the author of a number of nonfiction books, including *A Common Life: Four Generations of American Literary Friendship* and *The Reading Group Book*. A lifelong weather enthusiast, Laskin has published articles about the environment, travel, and family issues in numerous publications. He recently moved to the not-as-rainy-as-everyone-thinks climes of Seattle, Washington.